ENERGY TECHNOLOGY
Sources, Systems
and
Frontier Conversion

Titles of Related Interest

ARDEN, BURLEY & COLEMAN
1991 Solar World Congress, 4-vol set

BEI
Modern Power Station Practice, 3rd Edition

INSTITUTE OF ENERGY
Combustion and Emissions Control

INSTITUTE OF ENERGY
Ceramics in Energy Applications

MURRAY
Nuclear Energy, 4th Edition

TREBLE
Generating Electricity from the Sun

SAYIGH
Energy Conservation in Buildings

SAYIGH
Renewable Energy Technology and the Environment, 5-vol set

SAYIGH & McVEIGH
Solar Air Conditioning and Refrigeration

Journals of Related Interest
(free specimen copies gladly supplied on request)

Annals of Nuclear Energy
Applied Energy
Biomass and Bioenergy
Bioresource Technology
Energy - The International Journal
Energy Conversion and Management
Geothermics
International Journal of Hydrogen Energy
OPEC Review
Progress in Energy and Combustion Science
Progress in Nuclear Energy
Renewable Energy
Solar Energy

ENERGY TECHNOLOGY
Sources, Systems
and
Frontier Conversion

by

Tokio Ohta

Yokohama National University
Yokohama, Japan

PERGAMON

UK	Elsevier Science Ltd, The Boulevard, Langford Lane, Kidlington, Oxford OX5 1GB, UK
USA	Elsevier Science Inc., 660 White Plains Road, Tarrytown, New York 10591–5153, USA
JAPAN	Elsevier Science Japan, Tsunashima Building Annex, 3–20–12 Yushima, Bunkyo-ku, Tokyo 113, Japan

Copyright © 1994 Elsevier Science Ltd

All Rights Reserved. No part of this publication may be reproduced, stored in a retrieval system or transmitted in any form or by any means; electronic, electrostatic, magnetic tape, mechanical, photocopying, recording or otherwise, without permission in writing from the publishers.

First edition 1994

Library of Congress Cataloging in Publication Data

Ohta, Tokio
Energy technology: sources, systems, and frontier conversion/by
Tokio Ohta. – 1st ed.
p. cm.
Includes bibliographical references and index.
1. Power resources.   2. Power (Mechanics)   I. Title.
TJ163.2.O42   1994
621.042—dc20   94–11525

British Library Cataloguing in Publication Data

A catalogue record for this book is available from the British Library

ISBN 0 08 042132 6

In order to make this volume available as economically and as rapidly as possible it has been produced from the author's word processor disk. Every effort has been made to ensure an exact reproduction within the time available.

Printed and bound in Great Britain by Redwood Books, Trowbridge, Wiltshire.

Contents

Preface, ix

Chapter 1. Energy and Its Resource, 1
    1-1. Energy, 2
    1-2. Classification of Energy, 3
        (1) Mechanical Energy, 4
        (2) Energy of Electrical and Electromagnetic Systems, 5
        (3) Chemical Energy, 7
        (4) Heat (Thermal) Energy, 10
        (5) Photon Energy, 11
        (6) Some Technical Terms of Energy, 13
    1-3. Energy Resources, 16
        (1) Coal, 16
        (2) Petroleum, 18
        (3) Natural Gases, 20
        (4) Nuclear Energy, 21
    1-4. Concept of the Life Cycle Model, 27
        (1) Simple Estimation of Resource Life Time, 27
        (2) A Consideration of Life Cycle, 27
        (3) Some Numerical Conclusions, 29
        (4) Hubbert's Model, 31
    1-5. Natural Energy, 32
        (1) Solar Energy, 32
        (2) Hydro power, Wind power, and Ocean Energy, 35
        (3) Geothermal Energy, 43
        (4) Lunar Tide Energy, 44

Chapter 2. Energy Conversion, 45
    2-1. Matrix of Energy Conversions, 46
    2-2. Quasi Static Process, 50
        (1) Carnot's Cycle, 50
        (2) Electric Tray, 53
        (3) Thermo-Dielectric Conversion, 54
        (4) Conditions for the Possibility of Quasi-Static Conversions, 58
    2-3. Thermoelectric Type Conversions, 62
    2-4. Dynamic Conversion, 71
        (1) Electric Power Generator and Electromotor, 73
        (2) Magneto Hydro Dynamic Power Generation, 74
        (3) Resonant Conversion, 78
    2-5. Photon Energy Conversions, 80
        (1) External Photoelectric Effects, 81
        (2) Photovoltaic Power Generator (Solar Cell), 83

Chapter 3. Evaluation of Energy, 89
    3-1. Exergy, 91
        (1) Introduction of Exergy, 91
        (2) Energy Conversion and Exergy, 96
        (3) Exergy and Density Change, 98
        (4) Exergy of Substances (Chemical Exergy), 100

3-2. Cost Evaluation, 104
　　(1) Primary Energy, 104
　　(2) Secondary Energy, 105
　　(3) Political Cost, 108

Chapter 4. Energy Systems, 117
4-1. Energy Transfer System, 118
4-2. Examples of Energy Systems, 122
　　(1) Simplest Example, 122
　　(2) Solar-hydrogen Energy System, 123
　　(3) Multistage Conversion, 125
　　(4) Actual System, 126
　　(5) Hydrogen Energy System, 128
4-3. Energy Storage, 130
　　(1) General Survey: Storage and Transport, 130
　　(2) Practical Systems, 133
4-4. Energy Transport, 159
　　(1) General Survey, 159
　　(2) Costs of Transport, 160
　　(3) Pipe Line, 162
　　(4) Utility Power Lines, 164

Chapter 5. Frontier Energy Conversions, 169
5-1. General Survey, 170
5-2. Entropy Reduction Systems, 176
　　(1) Water-Splitting by Semiconductor Electrode, 176
　　(2) Solar Battery, 180
　　(3) Membrane, 181
　　(4) Catalyst, 185
5-3. Entropy Production Minimum Systems, 188
　　(1) Functionality Materials, 188
　　(2) Metal hydride, 191
　　(3) Superconductors, 196
5-4. Innovation of Systems, 201
　　(1) Introduction, 201
　　(2) Hydrogen Energy Systems, 202
　　(3) Refrigeration Survey and Alternatives, 206
　　(4) Frontiers and Their Heyday, 212

Appendices
A-1. Nomenclature and Units, 216
A-2. Physical Constants, 219
A-3. SI Base and SI Derived Units, 220
A-4. Conversion of Energy Units (A), 221
A-5. Conversion of Energy Units (B), 222
A-6. Energy Consumption of Some Manufacturing Industries, 223

References, 225

Indices, 227
　　I. Scientific and Technical Terms, 227

II. Institutes and Companies, 233
III. Scientists, 233

Preface

Given the significance of energy in the environmental problems of our world, it is urgently necessary that the leaders in civic and industrial societies have a more thorough understanding and appreciation of the existing states of energy systems and their related technologies. Such knowledge is all the more pertinent knowing that these energy disciplines are subject to rapid change as the emphasis shifts to solving the global environmental problems facing us.

So far man has been able to obtain enough fossil fuels to match his energy utilization requirements and so energy technologies have naturally concentrated on the following principles.

The first is to create a comfortable living space with air-conditioning, sound-proofing, and illumination. The second is to provide ample free time in daily life by accelerating vehicle traffic, completely provided with all necessary roads and facilities, and by the elongation of day-time by the illumination of night. The third is to save physical labor and even brain work in daily life. Because people wishes to avoid distress and embarrassment their endeavors often rely upon automation with computers. All of these trends have accelerated the demand for energies, *i.e.*, people today have been accustomed to life styles matching these technologies.

We are now entering a serious transition stage resulting from changes in the global environment; forefront in our minds being the rise of atmospheric temperature caused by the combustion of fossil fuels, and the concomitant apprehension for their depletion. Mitigation and adaptation strategies to deal with these issues will alter the course of subsequent technology development. This invites us to consider what are the frontiers of energy conversion technology.

We must enumerate, first of all, energy saving systems and their elementary technologies, the principles of which can be established by the law of entropy minimization, or exergy maximization, and also by establishing a cooperative network among primary, secondary, and utilization energy systems.

After the author's experience of studying and teaching for more than thirty years at Yokohama National University, The University of Tokyo, and Keio University, he had made up his mind to arrange the important parts of this field together for publication. This present work gleans the most significant sections of five books published earlier by the same author (listed 1 to 5 in the references) but with more emphasis given to new energy technology and 'hotter' topics.

The science of energy conversions and energy systems is essentially an interdisciplinary field embracing not only science and technology but also economics, sociology and politics.

A characteristic of the present book is that the important parts of a subject are repeatedly arranged throughout the book whenever related topics appear.

This book has been published concurrently with both the commemoration of my retirement as the president of Yokohama National University and with the 21st anniversary of the first oil crisis in 1973.

Lastly the author would like to express his cordial thanks to the late Dr. Takahashi Hidetoshi (Emeritus professor of physics, the University of Tokyo) and the late Dr. Akamatsu Hideo (Emeritus professor of chemistry, the University of Tokyo) for conducting Japanese scientists, one of whom is myself, to advanced energy conversion science, and to Dr. T. Nejat Veziroglu (professor of mechanical engineering, the University of Miami) for his encouraging friendship.

His thanks are also due to Dr. Philip Michael and Mr. Shigeharu Tanisho for reviewing the draft and preparing the final manuscript. He is also deeply indebted to Elsevier Science Ltd. for undertaking the publication of this book.

He would like to note with thanks that his wife Yohko helped him to prepare the manuscript.

March, 1994

Tokio Ohta

President, Yokohama
National University

Chapter 1
Energy and Its Resource

Water mill wheel in ancient Rome
(British Museum)

Chap. 1. Energy and Its Resource

1-1. Energy

The term "energy" was derived from the Greek "εργον" (ergon) meaning "work". Ergon signifies not only work but also vital power, vitality, and vigor. The term energy is still often used to describe such a vigorous state. The Greek term "αεργον" (aergon) used to express depressed state, yielded the name of the element "argon", one of the most inactive gases.

Energy is a generic term for the faculty, power or capacity for doing "work" possessed by a body or a system of bodies. If the term "work" in this definition implies physical work, then the energy is scientifically defined. Physical work is defined by the scalar product of a force vector f with the displacement vector r:

$$W = (f \cdot r). \qquad (1.1)$$

However, the exact meaning of energy was not established until the middle of the nineteenth century when kinetic energy was introduced. This is the easiest to understand. G.W. Leibniz (1647-1716) first proposed the concept that a moving body of mass m possesses "energy" by expressing

$$W = mv^2 \qquad (1.2)$$

instead of the term vital power used up to that time.

T. Young (1773-1829) subsequently published a book entitled "A Course of Lectures on Natural Philosophy" (1807) in which he stated "The term energy may be applied, with great propriety, to the product of mass or weight of a body: into the square of the number expressing its velocity". He recognized Leibniz's hypothesis very well.

A more advanced account of kinetic energy is due to G.G. Coriolis (1772-1843). His expression is

$$W = (1/2)mv^2. \qquad (1.3)$$

This was one of the most important steps in the establishment of the physical concept of energy.

During the development of modern physics, two remarkable innovations of the energy concept have been realized. One is the concept of energy quanta proposed by M.L. Planck (1858 - 1947) in 1900. He discovered that the energy of monochromatic light with frequency ν has an energy

$$W = nh\nu \qquad (1.4)$$

where n is an integer and h is Planck's constant (= 6.626×10^{-34} [Js]). As will be shown in the following section, we have five kinds of energy among which only light energy has a special characteristic different from the other kinds, that is to say, the energy magnitude of monochromatic light varies discontinuously from $nh\nu$ to (n $\pm$ m)$h\nu$, where m is also an integer.

Another important discovery was due to A. Einstein (1879 - 1955). In 1922, as a result of the theory of special relativity, an equivalency between mass and energy was proposed through the relationship

$$W = mc^2 \qquad (1.5)$$

where c is the velocity of light (= 2.99×10^8 [m/s]). This equation shows that the sum of mass and energy must be conserved in any energy conversion process, so that the sum of mass and energy in the universe remains constant forever. Equation (1.5) shows that a mass of 1 [kg] is equivalent to 9×10^{13} [kJ] which is the same as the heat energy generated by the combustion of 2.34 [million kl] of light oil.

Energy has been the most important and useful concept to the study of physics and chemistry since the start of modern science. Lagrange's equation in mechanics, the Hamiltonian operator in quantum mechanics, and Gibb's or Helmholtz's free energy in chemistry are typical examples. Natural phenomena occur along a path which minimizes the overall energy necessary for the process.

On the other hand, the concept of energy in society seems to be quite independent of scientific energy; it concerns the problems of petroleum, coal, natural gas, nuclear energy, and hydroelectric power, *etc.*, from an economic point of view. Recently global air pollution, including climatic change, became a new energy problem.

The primary solution to these problems is undoubtedly **energy conservation**, whose leading principle can be derived only from physics and chemistry. Associating the social energy problem with physics and chemistry is becoming more and more important but it does not always draw people's attention.

The frontiers in energy technology are thus derived from physical and chemical fundamentals. This book is written from such a standpoint, laying emphasis on the fundamentals as well as the advanced multifarious technologies.

1-2. Classification of Energy

Three classifications of energy are possible. The first is based upon the behavior, or state, that results from physical or chemical energy, such as electromagnetic energy, and will be discussed in detail in this section.

The second characterizes the energy systems used in society, that is to say, the sub-system of primary energy (energy resource) composed of fossil fuel, nuclear energy, and natural energy.

Secondary energy is the genetic name of energy that is processed (oil refining, conversion to electric energy, *etc.*) to be easier and cleaner to utilize. The main source of secondary energy is different in each country, *i.e.*, about 40% electrical power in Japan, more than 40% in U.S.A., and maybe about 10% in China, as of 1991. The share of electrical energy as a source of secondary energy has been increasing more and more. This trend shows that energy conversion to electrical energy is becoming increasingly important.

Utilization systems are multifarious and technological improvements in energy efficiency are urgently needed. It should be stressed that one must select the energy source which is most appropriate to the utilization purpose. Energy systems will be discussed in more detail in the next chapter.

The third kind of classification results from the primary energy source (resources). Energy resources are (1) **fossil fuel** (petroleum, coal, natural gas, tar sand, oil shale, and so on), (2) **nuclear energy** (nuclear fission, nuclear fusion), and (3) **natural energy** (biomass, hydropower, geothermal, ocean tide, ocean thermal, solar photovoltaic, solar heat, wind power, and so on). In 1992, the world's primary energy sources consisted of petroleum (38%), coal (30%), natural gas (20%), hydropower (7%), nuclear energy (5%), and other contribution (below 1%). However, these statistics do not include noncommercial fuels such as biomass in the underdeveloped countries. Each percentage is calculated based on its generated energy; among them the petroleum contribution occupied *ca.* 3.3 [billion kl] (3×10^{14} [kcal]), in 1991.

The scientific classification of energy is exactly possible by introducing physical parameters which describe complementally the force F acting upon a body or system and the corresponding displacement ξ. Here F and $d\xi$ do not necessarily imply a mechanical force or geometrical displacement. This classification can be expressed as

$$dW = (F \cdot d\xi) \qquad (1.6)$$

where dW is a differential of work with the dimension of energy (the unit is [J]). *Work is the same physical quantity as energy.* The parameters F and $d\xi$ satisfying Eq.(1.6) can be properly chosen according to the kind of energy and are described as "complemental". The generalized force and generalized displacement are often called the "intensive variable" and "extensive variable", respectively.

The classifications of energy are as follows.

(1) Mechanical Energy
(a) **Kinetic energy.** The parameters F and dr in Eq.(1.6) are force and displacement with their usual meanings and

$$dW = (F \cdot dr). \qquad (1.7)$$

If the generalized force is torque N, then we have

$$dW = Nd\vartheta, \tag{1.8}$$

where dq is an angle.

Equations (1.7) and (1.8) lead readily to other expressions such as

$$W = \frac{1}{2}mv^2, \tag{1.9}$$

$$W = \frac{1}{2}Iw^2, \tag{1.10}$$

respectively, where the Newtonian law of motion is applied. In Eq.(1.9) and (1.10), m, v, I, and w ($= d\theta/dt$) are the mass, velocity, moment of inertia, and rotation velocity, respectively.

The S.I. unit of mechanical energy is [J] (= [Nm]). If a force of magnitude 1 [N] is exerted on a body and it moves 1 [m] along the direction of force, then the force does 1 [J] of work and the energy received by the body is also 1 [J]. Another unit [erg] is defined by

$$10^7 \text{ [erg]} = 1 \text{ [J]} \tag{1.11}$$

A body with mass of 1 [kg] moving with velocity of 1 [m/s], possesses an energy of $\frac{1}{2}$ [J].

(b) Energy of fluid (gas and liquid). If fluid is subjected to a pressure p and changes its volume by dV, then work of

$$dW = pdV \tag{1.12}$$

is done. If the units of p and V are [Pa] (= [N/m²]) and [m³], respectively, then W is measured in [J].

The energy of dynamic and static fluid systems will be discussed later.

(2) Energy of Electrical and Electromagnetic Systems

(a) Electrostatic energy. If a body with an electric charge q is placed at the point of potential field (or voltage V) relative to another charge, then we may regard V and q as the intensive and extensive variables, respectively, and the work necessary to increase V by dV is

$$dW = qdV \tag{1.13}$$

The units of q and V are [C] (= [As]) and [V], respectively.

An electron placed in a potential field of 1 [V] has an energy of 1 [eV], where e ($=1.6 \times 10^{-19}$ [C]) is the electronic charge. This quantity is called an "electron volt" and is commonly used as an energy unit ($= 1.6 \times 10^{-19}$ [J]) in atomic physics.

(b) Energy of parallel plate condenser. Let us consider a parallel plate condenser with surface area S, thickness d, a dielectric constant ϵ for the material between the plates, a stored charge q, and voltage V. The energy of this condenser is

$$W_e = \int_0^V q \, dV = \int_0^V CV \, dV = \frac{1}{2} CV^2 \quad (1.14)$$

If the relationships $V = Ed$, and $C = \epsilon S/d$ are substituted in Eq.(1.14), then we have

$$W_e = \frac{1}{2} DESd \quad (1.15)$$

where $D \, (= \epsilon E)$ is the electric displacement. From this example, we may regard E and D as the complemental intensive and extensive variables, respectively, and Eq. (1.15) can be rewritten as

$$W_e = \int_0^E E \, dD , \quad (1.16)$$

which is the general expression for the energy of a dielectric system.

(c) Energy of electromagnetic induction. When an alternating current i is flowing in an electrical circuit with an induction L, the stored energy is calculated as follows. An electromotive force $-L di/dt$ is induced, therefore the work W_i, necessary to circulate an electric current i in time Δt is given by

$$dW_i = L \frac{di}{dt} i \, \Delta t$$

$$= \frac{d}{dt} (\frac{1}{2} Li^2) \, \Delta t, \quad (1.17)$$

then we have

$$W_i = \frac{1}{2} Li^2 \quad (1.18)$$

The unit of L is [H] (= [V·s/A] = [Wb/A]). The energy stored in a circuit with inductance $L = 1$ [H] is $\frac{1}{2}$ [J] if the electric current is 1 [A].

(d) Energy of electromagnetic wave. An electromagnetic wave is defined as a propagating electric wave combined with a magnetic wave. The oscillating field plane of the both waves are perpendicular to each other, *i.e.*, when the electric field E, and magnetic field H are in the yz-plane and in xz-plane respectively, the propagation direction is along the x-axis. We can describe the waves by

$$\left. \begin{array}{l} E_y(x,t) = f(x - ct) \\[2mm] H_z(x,t) = \sqrt{\dfrac{\epsilon_0}{\mu_0}} f(x - ct) \end{array} \right\} \quad (1.19)$$

where ϵ_0 and μ_0 are the dielectric constant and magnetic permeability, respectively, and $c \,(= 1/\sqrt{\epsilon_0 \mu_0} = 3 \times 10^8$ [m/s]) is the speed of light in a vacuum. The energy density is

$$W = W_e + W_m$$

$$= \frac{1}{2}(\epsilon_0 E^2 + \mu_0 H^2) \quad (1.20)$$

When equation (1.19) is substituted into Eq.(1.20), we have

$$W = \epsilon_0 c f^2 = E_y H_z . \quad (1.21)$$

Expression (1.21) can be generalized to define a new vector

$$\mathbf{S} = [\,\mathbf{E} \times \mathbf{H}\,] \quad (1.22)$$

which is called the Poynting's vector of the electromagnetic wave.

(e) Magnetic energy. The magnetic field H and the magnetic flux density $B \,(= \mu H)$ are analogous to the electric field E and the electric displacement $D \,(= \epsilon E)$. E and H are intensive variables while D and B are extensive variables, so that, following Eq.(1.15) the magnetic energy density is given as

$$W_m = \frac{1}{2} BH . \quad (1.23)$$

(3) Chemical Energy
(a) Cohesive energy of materials. Materials (solid, liquid, and gas) are composed of atoms or molecules which interact with each other while retaining their stability.

The interaction mechanism can be classified as follows.

8   Energy Technology

(i) *Molecular binding due to the van der Waal's potential between every two molecules of the same material.* Methane, petroleum, helium, oxygen, nitrogen, etc. are examples of molecular binding materials. Most fossil fuels belong to this type of material. Mankind's energy resources rely on the chemical energy stored in the fossil fuels for the following two reasons. The first is that a sufficient quantity of fossil fuels is readily available, and the second is that the release of the stored chemical binding energy can be readily accomplished, i.e., the ignition temperature is low enough to be attained without difficulty. The combustion mechanism consists of two steps, the first being vaporization from the liquid or solid state and the second being the transfer of electrons from the fuel to the oxidation material. Assuming the energy needed to make the transfer is ϵ_t, the ignition temperature T_i can be estimated by the equation

$$kT_i = \epsilon_t, \qquad (1.24)$$

where k (= 1.38×10^{-23} [J/K]) is the Boltzmann's constant. Most molecular binding materials have e_ts of the order of 1 [eV], so that T_i is around 1,000 -1,500 [K].

The oxidization reaction of fossil fuel (C_nH_m) is written

$$C_nH_m + (n + m/4)O_2 = nCO_2 + mH_2O + Q, \qquad (1.25)$$

where Q is the heat evolved from the cohesive energy difference between the (fuel + oxygen)-system and the product ($CO_2 + H_2O$)-system.

In the case of pentane (C_5H_{12}), we have $Q = 83.34$ [kcal/mol]; this shows that the transfer energy $\{Q/23.06 \times$ (number of transferred electrons)$\}$ is $\epsilon_r = 1.21$ [eV].

(ii) *Hydrogen binding originating from the proton exchange between component molecules.* Ice, KDP, etc., belong to this class of materials. Hydrogen binding energy is small compared to molecular binding but fuel materials are seldom found in this class.

(iii) *The metallic binding mechanism (the most complicated).* Several free electrons are commonly shared by many component ions to yield the cohesive force. Most metals, such as iron, copper, alloys, etc., are of this type of cohesive mechanism. The magnitude of the metallic binding energy is third largest.

(iv) *Ionic binding arising from the electrostatic Coulomb forces between positive and negative ions that compose a material.* Salt (NaCl) is the most common material of this type. Many ionic materials exist around us. The binding is strong and there no free electron exists. Therefore they have transparency and are strong insulators of electricity.

(v) *Covalent binding due to the exchange of electron pairs in the outermost electronic shells of the component molecules.* Organic materials such as ethylene (C_2H_4) possess this type of binding and the binding is diagrammatically written as

$$\begin{array}{c} \text{H} \\ \diagdown \\ \end{array} \!\!\! \text{C} = \text{C} \!\!\! \begin{array}{c} \text{H} \\ \diagup \\ \end{array}$$
$$\begin{array}{c} \diagup \\ \text{H} \end{array} \begin{array}{c} \diagdown \\ \text{H} \end{array}$$

where both C atoms jointly share two electrons. These organic materials are fossil fuels and/or refined chemical materials.

On the other hand, inorganic materials with covalent binding are very hard and are mostly realized in the elements of group IV of the periodic table, germanium, silicon, carbon, tin, and lead. Four valence electrons are shared jointly by the nearest four neighbouring germanium ions. Germanium and silicon are typical semiconductors widely used in electronics. The binding force of covalent inorganic crystals is the strongest among the five kinds of binding mechanisms. Diamond belongs to this category.

When Eq.(1.6) is applied to express the cohesive chemical energy W_c, the intensive variable and extensive variable are taken as the binding force affinity ([J/mol]) and quantity of the chemical material ([mol]), respectively.

(b) Density energy. The type of chemical energy originating from the density difference of chemical material in a fluid medium is called the "density" energy or the "density difference" energy. The density energy W is given by

$$dW_c = \mu dM \quad \text{with} \quad \mu = -T\frac{\partial S}{\partial M}, \tag{1.26}$$

where m and M are the chemical potential and the molar density, respectively, of the chemical system which typically consists of two liquids of different densities.

Huge amounts of undeveloped energy resources are possible by utilizing the chemical density energy of Na^+ and Cl^- between sea salt water and river fresh water. This system is sometimes referred to as "fresh-salt water electric generation."

The electrical potential difference available for practical use in this type of generating systems is shown in Fig. 1.1 and is given by

$$V_e = (f/F)(\mu_1 - \mu_2), \tag{1.27}$$

where F (= 96,485 [C/mol]) is Faraday's constant and f is a numerical coefficient.

Another important application of this type chemical energy is the well-known electric battery which will be discussed in detail later.

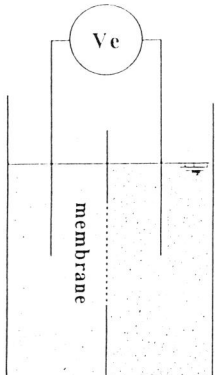

Fig. 1.1. Electrical potential difference of a two liquid system. Liquids with subscripts 1 and 2 have different densities to each other.

(4) Heat (thermal) energy

It was not such a long time ago that human beings became aware that "Heat is energy". Heat had long been regarded as an entity given the name "calorique", and often regarded as a substance or thing contained in or issuing from bodies ···", as described in the Oxford English Dictionary. R. Clausius, H. Helmholtz, and W. Thomson were the physicists who first established the thermodynamic system treating heat as a kind of energy in the latter half of nineteenth century. In 1865, R. Clausius found a new physical quantity, entropy (S), that is essential to the idea that "Heat is energy."

Heat energy can be expressed by, in reversible case

$$dQ = TdS, \tag{1.28}$$

where T and S are the intensive and extensive variables, respectively.

Taking entropy as defined by Eq. (1.28), we have, for any phenomena,

$$dS \geq dQ/T, \tag{1.29}$$

where the equality is valid only for reversible processes.

The concept of heat energy defined by Eq.(1.28) is very important not only in fundamental physics but also in energy application systems. L. Boltzmann discovered the relationship that is named after him

$$S = k \ln \Omega, \tag{1.30}$$

where Ω represents the thermodynamic probability. Equation (1.30) has a profound significance in that the energy and microscopic order of a physical system are closely related.

According to the principle of increasing entropy, the quantity of energy available in the natural world is constantly decreasing. Because entropy is produced successively and getting larger and larger the system temperature always tends to decrease. The most important point in energy conversion systems is that entropy production should be minimal and the entropy increment should also be minimal. Applications based upon this "entropy minimum leading principle" will be described in a later chapter.

(5) Photon energy (energy of light)

The energy of an electromagnetic wave, as explained in the section on electromagnetic energy, is the product formula of intensive and extensive variables in Eq.(1.23).

However, as the wavelength becomes shorter and shorter, the electromagnetic wave appears as a light wave whose energy cannot be expressed by the said formula. The concept of wave energy should therefore be replaced by photons. Photon energy is a quantum energy expressed by Eq.(1.4) which cannot be described as a traditional continuous variable systems. The temperature distribution of monochromatic light is $\exp(-nh\nu/kT)$, where the mean energy value is given by

$$\overline{W} = \sum_{n=0}^{\infty} nh\nu e^{-nh\nu/kT} \Big/ \sum_{n=0}^{\infty} e^{-nh\nu/kT}$$

$$= h\nu/[\exp(h\nu/kT) - 1]. \qquad (1.31)$$

Therefore the energy of white light is given by

$$\overline{W} = \frac{8\pi h}{c^3} \int_0^{\infty} \frac{\nu^3}{\exp(h\nu/kT) - 1} \qquad (1.32)$$

where we applied the density of states $g(n) = (8\pi/c^3)\nu^2 d\nu$ between ν and $\nu+d\nu$. Equation (1.32) is calculated to give

$$W = \sigma T^4 \qquad (1.33)$$

with

$$s = \frac{8p^5 k^4}{15ch^3}$$

$$= 5.174 \times 10^{-12} \text{ [J/K}^4\text{]}. \qquad (1.34)$$

Equation (1.33) is termed the "Stefan-Boltzmann's law" and σ, the "Stefan-Boltzmann's constant", respectively.

An empirical relationship between the maximum frequency ν_m and the temperature:

$$\nu_m/T = \text{constant} \qquad (1.35)$$

was theoretically proved by W. Wien in 1893. The constant is 2.9×10^{-3} [m·K], assuming the value of σ given by Eq.(1.34).

We must mention that the overwhelming solar energy incident on the surface of the earth has manufactured the biomass, fossil fuels, and all climatic energy.

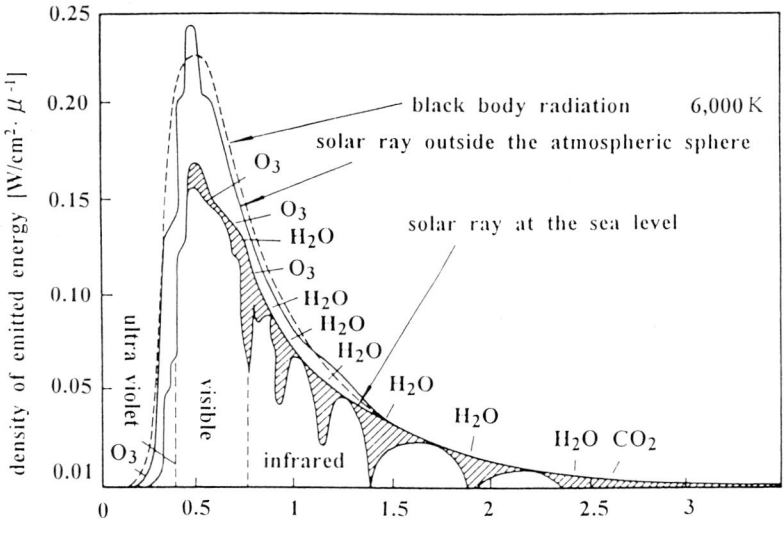

Fig. 1.2. Spectrum of solar energy. The denoted chemical materials indicate the absorbers.

The solar energy spectrum is shown in Fig.1.2. The shortest wavelength is about 0.1 [mm] and its radiant energy density is only about 0.1 [W/m²·mm]. Assuming that the surface temperature T_s of the sun is 6,000 [K] the energy distribution with wavelength is shown by the dotted curve in Fig.1.2. The solar energy density incident on the stratosphere is 1.4 [kW/m²]; it is reduced to below 1.0 [kW/m²] at the earth's surface, because many atmospheric materials (air, carbon dioxide, water vapor, dust, *etc.*) absorb or scatter the solar energy and the light energy is converted to heat. Much entropy is produced as the solar energy changes to heat.

Direct and positive utilization of solar energy is an important task. Some frontiers in this field will be described later.

(6) Some technical terms of energy

In addition to the classification of energy described so far, we must note some of the technical terms frequently used in science and technology.

(a) Potential energy. If the generalized displacement x has the dimension of length, the system's capacity for doing work is called "potential energy". Some examples are described below.

(i) *Potential energy in the earth's gravitational field.* If a body with mass m is placed at a position of height h_1 and then falls down to another position h_2, the work done by the body is

$$W = mg(h_1 - h_2) \tag{1.36}$$

where g (= 9.8 [m/s^2]) is the gravitational acceleration. In Eq.(1.36), mg and h correspond to F and r in Eq.(1.7), respectively.

The potential energy in a compressed gas is defined as $(p_1 - p_2)$, in the deformed state of an elastic body as $(1/2)Kx^2$ (K is the elastic constant and x is the displacement from the equilibrium position), and, in Coulomb's field, as $\pm q_1 q_2 / r$ (q_1 and q_2 are the electric charges or the magnetic charges and r is the distance between them).

(b) Bernoulli's theorem in fluid dynamics. For a fluid, the kinetic and potential energies are interrelated with the flow pressure through the relation

$$p_1 + \rho g h_1 + \frac{1}{2}\rho v_1^2 = p_2 + \rho g h_2 + \frac{1}{2}\rho v_2^2$$
$$= \text{constant} \tag{1.37}$$

where ρ is the fluid density. This relationship is called the "Bernoulli Equation." Equation (1.37) is fundamental to the study of any energy problem in fluid dynamics.

If wind with a velocity v passes through a windmill, the energy density done by the wind per unit area and per unit time is given by

$$E_0 = \frac{1}{2}\rho v^3 \ [\text{J/s·m}^2],$$

where ρ (= 1.225 [kg/m^3]) is the mean air density. It should be noted that the upper limit to the energy efficiency of a windmill is expressed by Betz's formula

$$\eta_m = E_m/E_0 = 16/27,$$

where E_m is the maximum used energy (p.36).

(c) **Energy of a living body.** One uses often the term "energy of a living body". There exists no such special kind of energy; it is just one kind of chemical energy. A living body maintains its life by the glycolytic pathway and respiration. The material produced during both processes is ATP (adenosine triphosphate), which is the energy resource of all living bodies. The reaction which generates this energy is reversible and is written as

$$\text{ATP} \quad \text{ADP} + \text{H}_3\text{PO}_4 + Q, \tag{1.38}$$

where ADP is adenosine diphosphate, which is ATP with one phosphoric acid missing, and Q (= 8 [kcal/mol]) is the generated energy. Animals can generate this energy in a very short time if necessary.

The energy consumption rate of an animal is expressed by

$$W_c = 4.1/M^{0.25} \quad [\text{J/kg·s}] \tag{1.39}$$

where M is the mass of the body. The energy consumption efficiency of a heavy animal is much better than that of a light animal. This field could be an important study subject in the future.

(d) **Nuclear energy.** Equation (1.5) indicates an equivalency between mass and energy. One calls the energy which is detained from mass conversion nuclear energy. However, we can also characterize nuclear energy according to the five kinds of energies discussed so far. Radioactive energy is an example of photon energy with very short wavelength.

(e) **Thermodynamic potential.** We shall concisely note some fundamentals of thermodynamics in order that the descriptions in the later chapters are easily understood.

We have two kinds of free energy, one of which is **Helmholtz's free energy** F defined by,

$$F = U - TS, \tag{1.40}$$

where U is the internal energy. Another is **Gibbs' free energy**, G, given by

$$G = F + pV$$

$$= U - TS + pV. \tag{1.41}$$

Enthalpy is defined by

$$H = U + pV. \tag{1.42}$$

The quantity of internal energy U, the free energies F and G, and the **enthalpy** H are called the thermodynamic potential or the thermodynamic function.
(f) Power. The work done in a unit time is called power. The unit of power is the watt, [W] (= [J/s]). An electric current of 1 [A] flowing between two points with a potential difference of 1 [V] does 1 [W] (= [J/s]) of work. In the practical field of energy engineering, the technical term "**horse power**" is often used. One horse power (1 [HP]) is defined as

$$\left. \begin{array}{l} 1 \text{ [HP]} = 0.7355 \text{ [kW] for French definition} \\ \\ 1 \text{ [HP]} = 0.7461 \text{ [kW] for British definition} \end{array} \right\} \quad (1.43)$$

Incidentally, we arrange the units of power in the following manner,

$$\left. \begin{array}{ll} 1[W] = 1[J/s] & 1[kW] = 1[kJ/s] \\ \\ 1[W] = 1[V \cdot A] & 1[kW \cdot h] = 3600[kJ] \end{array} \right\} \quad (1.44)$$

where 1 [h] = 3600 [s].

Let us consider lightning, an electrical discharge between a cloud and the earth, often generating an electric current of 10^5 [A] at a potential of 10^6 [V]. The power is 10^8 [kW], but the discharging time is usually less than 0.1 [s] so that the work done by the lightning is

$$W_l = 2.8 \times 10^3 \text{ [kW} \cdot \text{h]}$$
$$= 2.38 \times 10^6 \text{ [kcal]}. \quad (1.45)$$

The quantity W_l in Eq.(1.45) is about 40% of the chemical energy of one kiloliter of petroleum.

One should notice how the concept of power differs from that of force. We have

$$[W] = [N \cdot m/s], \quad (1.46)$$

i.e., the power is the product of force times velocity.

When ten tons of water per second fall from a 100 [m] high dam at a hydroelectric power station, the power is calculated as

$$P = 10 \text{ [t]} \times 9.8 \text{ [m/s}^2\text{]} \times 4.43 \text{ [m/s]}$$
$$= 434 \text{ [kW]}.$$

Consider another example of the combustion of fuel. If a gas burner consumes natural gas at the rate of 1 [m^3/s] to boil water, then the power is 10^4 [kcal/s]

16   Energy Technology

= 2,380 [kJ/s] = 2,380 [kW], where we assumed a heat quantity of 10^4 [J/m^3] for natural gas.

Hydrogen gas has a heat of combustion of 141.86 [kJ/g] at the higher limit and of 119.93 [kJ/g] at the lower limit. If a burner consumes 1 [m^3] of hydrogen gas in one second, then the power is at least 104 [kW]. Hydrogen gas is well suited to gas turbines because of its rapid combustion speed. Sakaide power station on Shikoku island has used a hydrogen rich (55%) gas to drive its gas turbine. The hydrogen is provided as a by-product from nearby chemical plants. Hydrogen is also a well known fuel for rocket engines at the second stage because the **thrust** (dimensions of [N], usually expressed by [kg] = 9.8 [N]) per weight of the fuel is very strong.

1-3. Energy Resources

(1) Coal

There exist four kinds of coal. They are Anthracite, Bituminous, Sub-bituminous, and Lignite. The first two are called "coal" in general and have been used widely in industrial and domestic applications. On the other hand, the latter two have relatively lower heat density and contain more sulphurous materials. They have been used in rather specialized industrial applications.

The confirmed reserve quantity, heat density, and annual production rate [t/Y] in 1983 are listed in Table 1.1.

Here we shall note the heat energy density. In Table 1.1, the average values of the HHV (**higher heating value**) and LHV (**lower heating value**) are shown. The difference between these values is essentially the latent heat of vaporization of water present in the exhaust products when the coal or other fuel is burned in dry air. Therefore, the difference is given by

$$HHV - LHV = 2,400 \ (M + 9H_2) \ [kJ/kg] \qquad (1.47)$$

where 2,400 [kJ/kg] is the latent heat of water vaporization, and M and H_2 are the moisture content and hydrogen mass fraction of the fuel. The HHV is usually determined experimentally, although a theoretical value can be obtained using Dulong's formula:

$$HHV = 33,950 \ C + 144,200 \ (H_2 - O_2/8 \ ) + 9,400 \ S \ [kJ/kg] \qquad (1.48)$$

where C, H_2, O_2, and S are the respective mass fractions of carbon, hydrogen, oxygen, and sulphur in the fuel.

The coal reserves in the world are very large. A simple calculation from the data in Table 1.1 gives a lifetime of 228.4 Y. In addition, the coal resources are fairly evenly distributed all over the world. However, underground coal mines have been largely unprofitable because of the heavy labor, countermeasures required for safety, *etc*..., and therefore surface coal mines are the mian suppliers today.

Table 1.1. Data of various coals

Classification	Confirmed reserve [10^8 t]	Production rate[1] [$\times 10^8$ t/Y]	Heat energy density [kJ/kg]	Heat energy density [kcal/kg]
Anthracite[C 94, $O_2$3, $H_2$3][2]	5,057	} 26.29	32,331	7,698
Bituminous[C 84, $O_2$10, $H_2$6]			32,564	7,753
Sub-bituminous[C 56, $O_2$37, $H_2$7]	2,220	} 5.56	29,308	6,978
Lignite[C 73, $O_2$21, $H_2$6]			25,586	6,092

Notes: 1) The production rates have increased slowly recently.
2) The numerical values show the percentages of the components.

The major difficulties associated with coal utilization include the following: the "air pollution" problem caused by the exhausted gas of coal combustion. More than three times the amount of SOx gas are exhausted compared to oil combustion. Another problem concerns the disposal of the remaining ashes.

Even in the age when air pollution was not such a concern as at present, coal was gradually replaced by oil and gas. The reason is that coal is not readily compatible with automatic combustion systems or with a pipeline transportation system. The economical disadvantages arising for these reasons were obvious.

If electric power is generated using coal combustion, the cost per [kW·h] is cheaper than that for oil generation only when no extra provisions are made for removing pollution. China is the world's largest coal user, accounting for 27 % of world production. More than 70 % of the country's energy is generated from coal. For example, the degree of cleanliness of the exhaust gas from coal combustion was 48 % and 25 % for U.S.A. and China, respectively, compared to Japan in 1990. If we can devise an inexpensive, but cleaner method of coal combustion, then the energy problem will be solved by the stability provided by the coal resources.

(2) Petroleum

There exist two kinds of world reserves for petroleum. One is the confirmed reserves (C.R.) and another is the ultimate reserves (U.R.). In 1985, British Petroleum Co. published a figure of $1,113 \times 10^8$ [kl] as C.R. As regards the U.R., most authorities published somewhat different values. For example, the value of U.R. was estimated to be $4,093 \times 10^8$ [kl] by the World Energy Conference of 1980, $2,731 \times 10^8$ [kl] by the 11th World Petroleum Conference, and $3,145 \times 10^8$ [kl] according to Japan Petroleum Society's data in 1986. As a general trend, we may estimate that the U.R. of petroleum is ca. $3,180 \times 10^8$ [kl], approximately 2×10^{12} [B] (barrel, 1 [B] = 0.159 [kl]).

Table 1.2. Distribution of confirmed oil reserves

Area	Oil Reserves (10^8 kl)	%
North America	54.5	4.9
Middle & South America & Europe	332.2	12.0
	42.3	3.8
Asia & Oceania	30.0	2.7
Africa	90.0	8.2
Russia	129.0	11.6
Middle East	631.0	56.8
Total	1,110	100

The production rate in 1985 was 32.8 $\times 10^8$ [kl/Y]. The magnitude of the production rate is presently not much different, but in 1979 (in the middle of the first oil crisis) it was 38.2 $\times 10^8$ [kl/Y], the maximum value in history.

We shall discuss the life cycle of petroleum in a later section.

The classifications and heat energy densities of several petroleum fuels are as follows. Kerosene: 40,850 [kJ/kg] (= 9,739 [kcal/kg]), burner oil: 42,680 [kJ/kg] (= 10,160 [kcal/kg]), heavy oil: 45,740 (= 10,900 [kcal/kg]), heavy oil (0.3% sulfur): 43,850 (= 10,440 [kcal/kg]). Because the density of kerosene is about 0.8 [kg/kl], its energy density per unit volume is about 12,160 [kcal/kl]. Gasoline, kerosene, light oil, and heavy oil are produced by the refining of crude oil.

The difficulties of petroleum fuels are as follows. The first problem is that the petroleum resources are unevenly distributed. The confirmed reserves in the world are estimated at about 1,110 $\times 10^8$ [kl] and these reserves are geographically distributed as shown in Table 1.2. It is obvious that about 57 % of the confirmed oil reserves exist in the Middle East, where the religious and racial confrontations are often violent and confusing. This situation sometimes prevents stable supply of oil and world energy crises are brought about.

Another difficulty associated with oil combustion is the global environmental problem. When oil is burned, as many kinds of gases are emitted as the number of its contained elements. The major gases are classified and discussed according to their pollution effects in a later chapter. Carbon dioxide (CO_2) has been recognized, for long time, as an non-noxious gas resulting from fossil fuel combustion. However, the huge amount of CO_2 gas accumulated year by year poses a serious problem to the environment.

As described before, the consumption rates of oil and coal in the middle of the 1980s were about 33 $\times 10^8$ [kl] and 32 $\times 10^8$ [t], respectively. The exhaust gases from the combustion of these fuels contain CO_2, the total amount of which was estimated to be 213 $\times 10^8$ [t], where 96 $\times 10^8$ [t] and 117$\times 10^8$ [t] come from oil and coal, respectively.

Carbon dioxide, which is well known as giving rise to the greenhouse effect, has been accumulating year by year. Consequently, the global average temperature is predicted to rise, resulting in many ecological problems. The same problem also exists to some extent in the case of methane gas. We shall further discuss this subject in detail in a later chapter. Nervertheless the demands on the oil supply have increased not only because of its economic merits as an energy source but also as an energy carrier and energy reservoir. In addition, many manufactured goods are made of oil products; these account for about 10 % of the annual oil production. The later half of the twentteth century and perhaps a few decades near the beginning of the twenty-first century are or will be called the "**Oil Age**".

(3) Natural gases

(a) Petroleum gases. Natural fuel gases can be classified into two kinds according to whether they are produced coincident with crude oil in oil fields or are produced independent of oil. Some gases which are produced along with crude oil also exist everywhere on the earth's surface. Methane is a typical gas of this kind. Methane, ethane, propane, *n*- or *iso*-butane, and other gases are known as "petroleum gas", because they are yielded along with crude oil. The properties of these gases are shown in Table 1.3. They are in the liquid state at the high pressures that exist underground, but become gaseous at atmospheric pressure and ambient temperature. Their boiling temperatures are below 0°C. As shown in Table 1.3, natural gas - C_nH_m - has a lower boiling temperature and higher heat density per mol as the ratio n/m becomes larger. Petroleum with the ratio value of $1.0 \leq m/n \leq 2.2$ is in liquid state at room temperature and atmospheric pressure.

Ethane, propane, and butane are usually used in LPG (liquefied petroleum gas, or often propane gas) as a substitute for gasoline or kerosene. For example, the total number of LPG automobiles in the world is about 3.65×10^6, ten percent of which are driven in Japan, where more than 99% of the taxis are driven by LPG.

These gases are very cheap at the oil fields, hence most of them are burned out on site. The majority of LPG imports are due to Japan, where many homes use LPG instead of city gas. The numbers of Japanese homes using LPG and city gas are nearly equal.

The disadvantage of using LPG is that the raw petroleum gases are difficult to obtain, not enough to meet demands, whenever OPEC (Organization of Petroleum Exporting Countries) undertakes a policy of reducing oil output.

(b) Methane gas. The confirmed world reserve of methane gas is estimated at about 100×10^{12} [m^3] which is equivalent to $1,050 \times 10^8$ [kl] of oil, a quantity roughly equal to the confirmed oil reserves. These gas reserves are also a little more evenly distributed as shown in Table 1.4. Furthermore, because the quantity of CO_2 produced from methane combustion is about half of that from oil at equivalent heat emission, it can be said that natural gas is gentler to the global environment than oil.

Table 1.3. Examples of natural gases (C_nH_m)

m/n	Name	Boiling temperature [°C]	Heat density [kcal/m^3]
4	methane (CH$_4$)	-161.49	9,500
3	ethane (C$_2$H$_6$)	-89.0	16,640
2.7	propane (C$_3$H$_8$)	-42.1	24,300
2.5	n-butane (C$_4$H$_{10}$)	-0.5	30,684

Table 1.4. Share of gas reserves
(*Oil & Gas Journal*, 1985)

Area	Share %
CIS (Russia *et al.*)	44.4
Middle East	24.6
North America	8.5
Europe	6.5
Africa	5.7
Middle & South America	5.4
Asia & Oceania	4.9

The World Gas Conference of 1985 has estimated the ultimate reserves of methane gas to be 300×10^{12} [m^3], therefore about 200×10^{12} [m^3] of reserves are not yet discovered. Most of them are believed to exist in the ground beneath Siberia, contained in the frozen earth.

The production rate of methane gas was 1.7×10^{12} [m^3/Y] in 1985 (not so much different from today), so that a simple estimation of the lifetime is 58.8 [Y]. According to the more exact Hubbert's life cycle model, which will be discussed in a later section, by 2070 the production rate, assuming an initial 3 % increase every year, will eventually drop to the level of 1960, *i.e.*, $4{,}000 \times 10^8$ [m^3/Y] as the oil reserves are exhausted. This inference shows that the lifetimes of both oil and gas are nearly same.

LNG is liquefied natural gas. Almost all of the manufactured quantities are imported by Japan. A major investment is necessary to construct a liquefaction facility and therefore it is impossible economically to have LNG plants in gas fields with few reserves. To overcome this difficulty, a shipbuilding plan for constructing a LNG tanker with an onboard liquefaction facility was concluded in Japan, in 1992. Such a tanker can access even small scale gas fields and manufacture LNG to transport to Japan.

(4) Nuclear energy
(a) Source of nuclear energy. According to Einstein's special theory of relativity, the relationship

$$W = mc^2 \tag{1.5}$$

22 Energy Technology

is known. In any chemical reaction, it is impossible to confirm Eq. (1.5) because the mass difference before and after the reaction is too small to measure. However, the mass difference, also called the mass defect, is appreciable when nuclear reactions occur. The mass of an atomic nucleus is less than the total mass of the individual particles (nucleons) that comprise it. This mass defect is what keeps the nucleus together. To break a nucleus into its individual nucleons, an amount of energy, at least equivalent to the mass defect, is necessary. The mass defect is usually estimated by

$$\delta m = Zm_H + (A - Z) m_n - \text{atomic mass} \qquad (1.49)$$

where atomic mass is used instead of nuclear mass, A and Z are the atomic mass number (total number of nucleons in the atom) and the atomic number, respectively, m_H and m_n are the masses of proton and neutron, respectively. Atomic masses are usually described in [amu] (atomic mass unit, approximately equal to proton or neutron mass) which is defined by

$$1 \text{ [amu]} = 1 / (\text{Avogadro's number in [g]})$$

$$= 1.66 \times 10^{-24} \text{ [g]}. \qquad (1.50)$$

In our unit system, the mass of carbon 12, ^{12}C, is exactly taken to be 12.00000 [amu], but in another system the standard is sometimes chosen to be the oxygen 16, so that the mass of ^{16}O is 16.00000. Using the values:

$$m_H = 1.007825 \text{ and } m_n = 1.008655,$$

we have

$$\delta m = 1.007825 Z + 1.008665 (A - Z) - \text{atomic mass} \qquad (1.51)$$

The mass defect increases as the atomic mass number increases because there exist more nucleons. The energy equivalent of the mass defect is called the total binding energy. It should be noted that

$$1[\text{amu}] = 931 \text{ [MeV]} \qquad (1.52)$$

The average mass defect of binding energy per nucleon increases steeply before reaching a maximum value. Then it slowly decreases as the atomic mass number increases. Figure 1.3 shows the specific binding energy, defined by $\delta m/A$, vs mass number. The elements, $^{56}_{26}Fe$, $^{59}_{28}Ni$, $^{54}_{27}Co$, and $^{62}_{30}Zn$, all with a mass number around 60, are the most stable. On the other hand, the elements, $^{1}_{1}H$ and $^{238}_{92}U$ which have least binding energy are the easiest to break down.

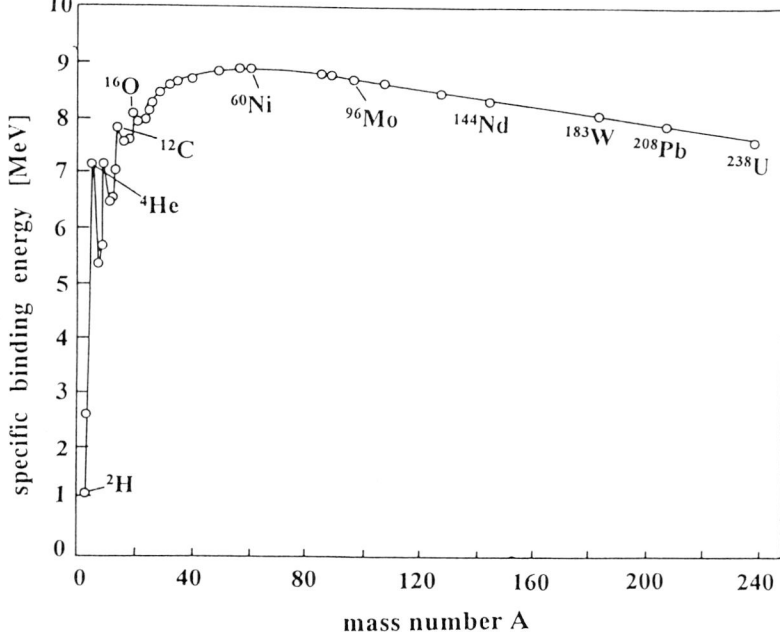

Fig. 1.3. Specific binding energy *vs* mass number for the natural elements.

(i) *Nuclear fission.* O. Hahn and F. Strassmann first discovered, in 1938, the release of nuclear binding energy by irradiating uranium 235, $^{235}_{92}U$ with neutrons. After absorbing neutrons uranium 235 becomes unstable and splits according to the reaction:

$$\left.\begin{array}{c}
^{235}_{92}U + ^{1}_{0}n \rightarrow {}^{94}_{38}Sr + ^{140}_{54}Xe + 2\,^{1}_{0}n \\[6pt]
^{94}_{38}Sr \xrightarrow{\beta} {}^{94}_{39}Y \xrightarrow{\beta} {}^{94}_{46}Zr \text{ (stable)} \\[6pt]
^{140}_{54}Xe \xrightarrow{\beta} {}^{140}_{55}Cs \xrightarrow{\beta} {}^{140}_{56}Ba \\[6pt]
\xrightarrow{\beta} {}^{140}_{57}La \xrightarrow{\beta} {}^{140}_{58}Ce \text{ (stable)}
\end{array}\right\} \quad (1.53)$$

where β denotes a β-decay. The total atomic mass of the uranium and the neutron is 236.0526 [amu] and that of the total reaction products is 93.9063(Zr) + 139.9053(Ce) + 2.0173(2n) = 235.8289 [amu], therefore the release of energy is

0.2237 [amu] (= 236.052 - 235.8289); this is equivalent to 208.4 [MeV] (1 M = 10^6). Including the energy of β-decays, we have more than 220 [MeV] of energy from the nuclear fission of one uranium 235 atom.

We should also note the two neutrons produced during the reaction (1.44). By applying these neutrons to another uranium atom a nuclear chain reaction is possible. Thus we have illustrated the principle of a nuclear reactor.

There are 14 isotopes of uranium whose mass numbers range from 227 to 240. Uranium with mass number 235 is available for the reaction listed in Eq.(1.44). Uranium resources exist as UO_2, [$(U,Ca,Fe,Th,Y)_3Ti_{15}O_{16}$], ($CaO \cdot 2UO_3P_2O_5 \cdot 8H_2O$), ($CuO \cdot 2UO_3P_2O_5 \cdot 8H_2O$), and ($K_2O \cdot 2UO_3V_2O_5 \cdot 8H_2O$) where the uranium is contained as oxides.

The percentages of uranium oxide and of ^{235}U in these compounds are 0.1-0.3 % and 0.7 %, respectively. Most of the remainder is ^{238}U which cannot sustain the nuclear reaction.

The critical mass of $^{235}_{92}U$ with 90 % purity is about 1 [kg], *i.e.*, most of the neutrons produced by this reaction are absorbed by other nuclei and give rise to chain reactions. To control these chain reactions, control bars made of cadmium are used. Cadmium absorbs the neutrons to limit the number of potential chain reactions. The neutrons emitted from the uranium nucleus are so fast (average energy is about 2 [MeV]) that they can hardly be absorbed even by the cadmium bars and hence a moderator is also necessary.

Nuclear reactors are classified by the kinds of moderators used and also by their coolants. The moderators are light-water, heavy-water, graphite, beryllium, *etc.*, and the coolants are water, gases and liquid metals. Reactors are classified according to the kinds of moderator, the type of coolant, and combinations of them, for examples, "heavy water reactor", "gaseous coolant reactor", and so on.

Reactors are also classified by their process of heat generation. One uses water vapor under atmospheric pressure and another the vapor from a pressurized container. They are called boiling water- and the pressurized water-reactors (BWR and PWR), respectively.

Next, we shall briefly mention breeder reactors. The following nuclear reactions occur if uranium 238 is placed around the uranium 235 in the reactor. The breeding reactions are

$$\left. \begin{array}{l} {}^{238}_{92}U + {}^{1}_{0}n \xrightarrow{\gamma} {}^{239}_{92}U \\[6pt] {}^{239}_{92}U \xrightarrow{\gamma} {}^{239}_{93}Np + {}^{0}_{-1}\beta \ (t = 23.4 \text{ [min]}) \\[6pt] {}^{239}_{93}Np \xrightarrow{\gamma} {}^{239}_{94}Pu + {}^{0}_{-1}\beta \ (t = 2.35 \text{[D]}) \\[6pt] {}^{239}_{94}Pu \xrightarrow{\gamma} {}^{235}_{92}Pu + {}^{4}_{2}\alpha \ (t = 24{,}400 \text{ [Y]}) \end{array} \right\} \quad (1.54)$$

where γ means γ-decay.

One should notice that $^{239}_{94}$Pu has a very long half-life and is able to sustain nuclear fission by absorbing neutrons. The uranium 238 itself is not fissionable but serves to generate the plutonium 239 which is called the generating material. A nuclear reactor where more plutonium 239 is produced than the quantity of uranium 235 placed at the center of the reactor is called "a breeder reactor". Once breeder reactor systems are completely successful, we will not need to worry about our energy resources.

(ii) *Nuclear fusion.* The masses of the proton and the neutron are 1.007825 [amu] and 1.008655 [amu], respectively, so that two of each nucleon comprises 4.03296 [amu]. On the other hand, the mass of helium 4 is only 4.00260 [amu]; the difference 0.03036 [amu] will be released from a nuclear reaction that combines two pairs of proton-neutron to produce one helium atom. This mass defect is equivalent to 28.265 [MeV], and a huge quantity of energy will be released. For example, 1 [kg] of this deuteron pair will yield 2.728 ×10^{12} [kJ] following the nuclear fusion reaction (1 [MeV] = 1.602 ×10^{-16} [kJ]).

Exact calculations which account for the mass difference between the nuclear reactants and the products are shown in Table 1.5 for some possible fusion reactions. The largest energy release is 18.3 [MeV] in the $^3_2$He + $^2_1$H reaction and the minimum threshold energy is 10 [keV] for the $^3_1$H + $^2_1$H reaction, where the ignition temperature is 1.2 ×10^8 [K]. At such high temperature, all of the electrons are stripped from the nuclei and reactants are in the plasma state. In the plasma state, all component elements are in the ionic state and the coulomb repulsion between them is strong enough to prevent fusion. Considerable research has been devoted to achieving controllable nuclear fusion, but it has not succeeded yet.

Table 1.5. Nuclear fusion reactions and their ignition conditions

Reaction		Ignition Energy	Ignition Temperature
$^2_1$H + $^2_1$H → $^3_2$He + $^1_0$n + 3.26 MeV		50 keV	5.8 × 10^8 K
$^2_1$H + $^2_1$H → $^3_1$H + $^1_1$H + 4.03 MeV		50 keV	5.8 × 10^8 K
$^3_1$H + $^2_1$H → $^4_2$He + $^1_0$n + 17.4 MeV		10 KeV	1.2 × 10^8 K
$^3_2$He + $^2_1$H → $^4_2$He + $^1_1$H + 18.3 MeV		100 KeV	1.2 × 10^8 K
$^6_3$Li + $^1_1$H → $^3_2$He + $^4_2$He + 4.0 MeV		200 KeV	2.3 × 10^9 K

To achieve nuclear fusion for reactants in the plasma state, **Lawson's condition** should be satisfied, which at $T = 10^8$ [K] is written as

$$N\tau = 10^{21} \ [\text{s/m}^3], \qquad (1.55)$$

where N and τ are the ion density and the confinement time, respectively.

The hydrogen bomb is an explosive release of nuclear fusion energy ignited by a conventional atomic bomb (nuclear fission).

(b) Resources. The resource for nuclear fission energy is uranium 235 which can be manufactured by condensing natural uranium using the centrifugal method. The total amount of the uranium in the world is considerable because even seawater contains uranium with a density of 3.3×10^{-6} [kg/kl]. However, the investment for a condensation facility for seawater is too great to be a practical application. The amount of uranium resources is a function of the price. The confirmed reserves (C.R.) which can be provided with the price under U.S.$ 80/kg is about 3.1×10^6 [t] and with the price under U.S.$130/kg is 5.2×10^6 [t] in C.R. and 13.27×10^6 [t] in U.R. The price of U.S.$ 80/kg is considered equivalent to oil when both fuels are applied to generating electric power in a power station.

The life cycle model of the uranium fuel is not so simple to evaluate as fossil fuels because an explicit form of the function for the production rate and C.R. with the time and the price are not definite, however, we may infer that the U.R. which can compete economically with oil is comparable or less than that of oil.

The reactants for various nuclear fusion reactions are shown in Table 1.5. They are deuterium ($_1^2\text{H}$), tritium ($_1^3\text{H}$ or $_1^3\text{T}$), helium 3 ($_2^3\text{He}$), and lithium ($_3^6\text{Li}$). The density of atomic deuterium in seawater is 0.015 %. Tritium does not exist in nature and is usually produced by the reaction:

$$_3^6\text{Li} + {_0^1}\text{n} \rightarrow {_2^4}\text{He} + {_1^3}\text{H} + 4.8 \ [\text{MeV}]. \qquad (1.56)$$

The world's underground lithium reserves are estimated at $8 - 9 \times 10^6$ [t]. Lithium also exists in seawater at a density of 1.7×10^{-4} [kg/m^3]. Helium 3 ($_2^3\text{He}$) comprises only 1.3×10^{-4} % of the natural helium gas supply and is very expensive to separate.

Considering the quantity of reactants in the world, one should take seawater as the principal source of deuterium. In order to condense large quantities of seawater, we need a large scale facility accompanied by a very large investment. It is most useful to produce deuterium by the electrolysis of common water. This process is aided by the fact that the electrolysis voltage of heavy water is a little higher than that of light water. The boiling temperature of heavy water is 101.4°C, while its melting temperature is 3.82°C, both of which are higher than those of light water. These properties also apply to the manufacture of heavy water.

Deuterium can be separated from heavy water by applying a chemical reaction using sodium. It is then a matter of course to apply centrifugal separation.

(c) **Cold fusion.**[9, 29] S. Pons and M. Fleishmann (University of Utah) and also S. Jones (Brigham Young University) reported, on 23rd of March 1989, that an extraordinary evolution of heat and emission of neutrons as observed during a heavy water electrolysis process where the anode was made of platinum and the cathode, where deuterium was absorbed, of palladium. They issued a comment that this phenomenon must be a kind of nuclear fusion, and all the scientists in the world took notice and called it "cold fusion".

This phenomenon could seldom be reproduced in the same experimental system but has been reported successively by several researchers around the world. A theory to explain it has not been given yet and many scientists doubt that there ever will be. We have no confirmed conclusion at present.

1-4. The Concept of the Life Cycle Model

(1) Simple estimation of resource life time

The production rates and global reserves of every energy resource are usually known as a function of time elapsed, for example, the data of petroleum for 1985 indicates a production rate of $dM/dt = 3.48 \times 10^9$ [kl/Y] (dM is the quantity produced during the time dt) and total global reserves of $M_0 = 3.18 \times 10^{11}$ [kl]. Assuming a constant consumption rate, the total life of the reserve can be estimated.

If petroleum is consumed at the same constant rate, then it will be exhausted after the time $t_\infty = M_0/(dM/dt)$, 91 years after 1985.

However, we know that the situation is not so simple. The production rate has been sensitively influenced by the global economy and the demands of importing and exporting countries. Since petroleum first began to play an important role in the economy the production rate has been increasing on average and is rather uniform at present. On the other hand, it is properly anticipated that the rate will decrease after a time, when about half of the reserve amount is consumed. In order to express this trend, and to study more exactly the resource life cycle, we shall study a model using the assumptions below.

(2) A consideration of life cycle

To derive numerical data, such as the lifetime or half lifetime of a resource, we set up a mathematical model using the following assumptions, and taking petroleum as an example.

(i) *The total initial amount of the global reserves is constant and denoted by M_0.*

(ii) *The production rate (equal to the consumption rate of the reserves) increases rapidly until the remaining amount reaches half of the total initial amount of the reserve.* The half lifetime τ is defined as the time when the remaining amount reaches half of the initial values, so that we have

$$M(\tau) = M(0)/2 \tag{1.57}$$

(iii) *The production rate decreases rapidly after the half lifetime.*
(iv) *Both the increasing and the decreasing consumption rates are proportional to the remaining amount.*

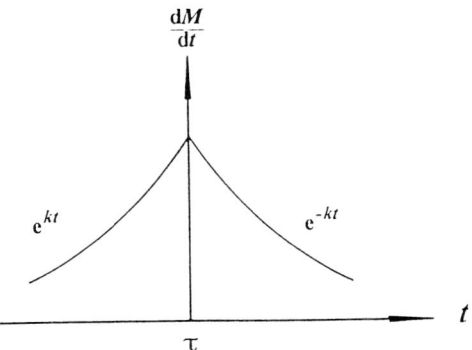

Fig. 1.4. Production rate of oil *vs* time.
It increases exponentially until half liftime and then decreases exponetially after half lifetime.

The amount remaining at the time t is denoted by $M_+(t)$, before the half life and by $M_-(t)$ after the half life, and we can write

$$M(0) = M_+(0) + M_-(\tau) \tag{1.58}$$

$$M_+(0) = M_-(\tau) = M(\tau) = M(0)/2, \tag{1.59}$$

$$M(t) = [\, M_+(0) - \int_0^t \left(\frac{dM_+}{dt}\right) dt \,]$$

$$+ [\, M_-(t) - \int_\tau^t \left(\frac{dM_-}{dt}\right) dt \,]. \tag{1.60}$$

Assuming two different time constants k_1 and k_2 before and after the half lifetime t are defined by

$$\frac{dM_+}{dt} = k_1 M_+ \quad \text{and} \quad \frac{dM_-}{dt} = -k_2 M_- \tag{1.61}$$

we get (refer to Figs. 1.4 and 1.5)

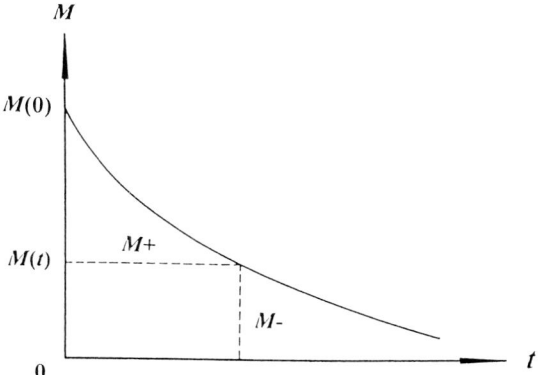

Fig. 1.5. Oil resource decaying with time. The decaying rate is different after the half lifetime τ.

$$M_+ = M(0) \exp(k_1 t) \quad \text{for} \quad 0 \leq t \leq \tau \tag{1.62}$$

$$M_- = M(\tau) \exp(-k_2 t) \quad \text{for} \quad \tau \leq t \leq \infty. \tag{1.63}$$

Substituting Eqs. (1.62) and (1.63) into Eq.(1.60), we get the expression for the total amount remaining at a time t, i.e.,

$$M(t) = M_+(t) + M_-(t), \tag{1.64}$$

with

$$M_+(t) = M(0)[2 - \exp(k_1 t)], \quad \text{for} \quad 0 \leq t \leq \tau \tag{1.65}$$

and

$$M_-(t) = M(\tau) \exp(-k_2\{t-\tau\}), \quad \text{for} \quad \tau \leq t \leq \infty. \tag{1.66}$$

We confirm that $M_+(0) = M(0)$ at $t = 0$ and $M_-(t) = M(\tau)$ at $t = \tau$ by using Eq.(1.65) and Eq.(1.66), respectively.

(3) Some numerical conclusions

The time constant k_1 is emperically determined by substituting $(dM_+/dt) = 3.48 \times 10^9$ [kl/Y] and $M_+ = 1.59 \times 10^{11}$ [kl] into Eq.(1.61), to obtain the value

$$k_1 = 1.38 \times 10^{-2} \; [1/Y]. \tag{1.67}$$

Now, we can estimate the half lifetime τ using Eq. (1.65). The remaining amount at $t = \tau$ is half of the total quantity, therefore we have

$$\tau = \ln(3/2)/k_1$$

$$= 0.4054/k_1. \qquad (1.68)$$

If the k_1 value given by Eq.(1.67) is substituted into Eq.(1.68), we get

$$\tau = 29 \text{ [Y]}. \qquad (1.69)$$

The above calculation shows that the rapid production rate expressed by Eq.(1.61) results in a very short half lifetime, but the time at which the resources will be exhausted is much longer than $2\tau = 58$ [Y].

As an example, we shall seek the time when the petroleum resources reach 10 % of their initial value. Equation (1.65) shows that 50% of the resources were consumed at time t, and if 40% of the remainder was consumed, we have

$$M_- = M(\tau)/10$$

and

$$t\,(90\%) = 37 + 0.916/k_2. \qquad (1.70)$$

If k_2 is assumed to equal k_1, we have

$$t\,(90\,\%) = 37 + 84$$

$$= 121 \text{ [Y]}. \qquad (1.71)$$

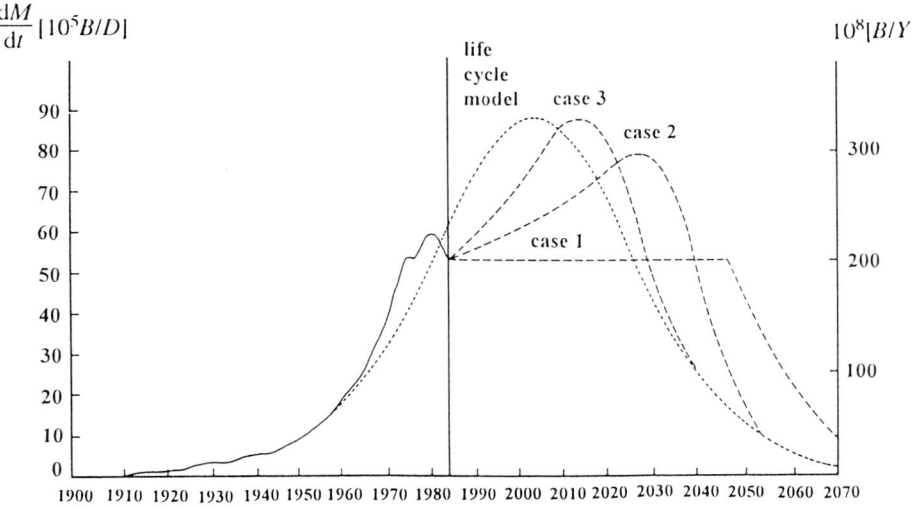

Fig. 1.6. Hubbert's model for petroleum.
After the original paper, the unit 1 [B] = 0.159 [kl] is used.

The assumed production rate $|dM_\pm/dt|$ is proportional to the time constants $k_{1,2}$ but we may infer that k_2 will be smaller than k_1 so that the remaining resources will be consumed more and more slowly as the remaining quantity decreases. Therefore we may assume that both k_1 and k_2 will depend on time.

(4) Hubbert's Model

M.K. Hubbert (U.S.A.) proposed a life cycle model for petroleum in 1956. His model is well known and called "Hubbert's Model". The model is based upon the following three assumptions:

(i) *The total initial petroleum reserve is 2×10^{12} [barrel] (1 [barrel] $= 0.159$ [kl]), i.e., $3,180 \times 10^8$ [kl]*.

(ii) *The production rate increases to a maximum and then saturates for a short time before decreasing.*

(iii) *The behavior of the production rate is assumed by an extension of empirical data and is assumed to be symmetrical with respect to the time when the production rate becomes maximum.* To complete the model, we extended the empirical data taken from 1910 to 1955 by Hubbert to 1980. Figure 1.6 is plotted using Hubbert's model. In addition to his life cycle three other cases are also shown.

(i) Case 1: The production rate is constant until 2045 and then decays. (ii) Case 2: The production rate increases by 1 % every year. (iii) Case 3: The production rate increases by 2 % every year. This is most probable. Under this assumption, the

Table 1.6. Resources capacity of natural energy in the world

Classification	Capacity	Kind of energy
Solar energy (absorbed in the atmosphere)	0.4×10^{14} kW*	Photon and heat energies
Hydro power	3.0×10^9 kW	Kinetic energy
Wind power	9.7×10^9 kW**	Kinetic energy
Geothermal	4.0×10^{17} kJ	Kinetic energy
Tide energy	6.7×10^7 kW	Kinetic energy

*Refer to Fig. 1.4 and Ref.(11).
**This values is the U.S.A only.

production rate becomes maximum at 2010 where it is about 50 % larger compared with that in 1985, *i.e.*, 52.2 x10^8 [kl/Y]. One can infer from Fig. 1.6 that petroleum will be exhausted by about 2062.

1-5. Natural Energy

Natural energy indicates the energy that drives, or activates natural phenomena. Similar to the sources of energy, we can classify natural energies into three categories. The first is solar energy which drives more than 99.9 % of the activating phenomena on the earth's surface. The second is the earth itself which consists of high temperature crusts where hot magmas generate water vapor. The first geothermal power plant was established in 1904, in Italy. The third energy source is the lunar attractive force exerted on the oceans. The energy utilization of ocean tides has a long history, since 1966 in France.

The capacity and classification of natural energy sources are shown in Table 1.6, where it should be noted that both hydro power and wind power originate from solar energy.

(1) Solar energy
(a) Resource. The nuclear fusion reactions in the sun yield a huge amount of energy which is estimated at 3.47 x10^{24} [kJ] per unit time. This value is comparable with the revolving kinetic energy of the moon around the earth. Of this huge amount of emitted energy, only a small part, 5 x10^{-11}, is irradiated onto the earth's surface. The incident solar energy is distributed into many branches as shown in Fig. 1.7. Solar energy is said to be clean and undepletable. It is true that solar energy is harmless to living bodies on the earth's surface because the harmful short wavelength ultra-violet rays are absorbed by the stratospheric ozone layers and weakened by the air and moisture in the atmosphere. However, if the ozone layers are destroyed by artificial chemical substances such as flon CFC(Chlorofluorocarbon), living bodies will be subject to serious effects and human beings especially will be prone to skin cancer.

(b) Utilization. Solar energy activates the atmosphere thus generating climatic phenomena, but the balance of the energy is absorbed by molecules of the materials on the earth and converted into heat at low temperature. This is an example of the entropy increasing process of nature. It is reasonable to plan to actively utilize the sun's photon- and high temperature heat-energies before they decay to produce entropy.

Figure 1.7 also shows the distribution of the incident solar energy at the earth's surface. The artificial utilizations of solar energy are also listed there. Two classifications are possible, the first is due to photon energy, the quality of which is much higher than that of heat. However, photon energy is readily converted to heat as soon as it is absorbed.

The most advanced technology of photon utilization is the solar cell, to which the photovoltaic effect of semiconductors is applied. The highest efficiency achieved of a solar cell with tandem type use of GaAs - GaSb semiconductors is 35.6%

Energy and Its Resource   33

Fig.1.7. Distribution of the solar energy onto the earth's surface and utilization.
 (1) *Photon energy means the energy of short wave which is immediately converted to heat when absorbed by molecules.
 (2) Utilizations in brackets are the main applications.

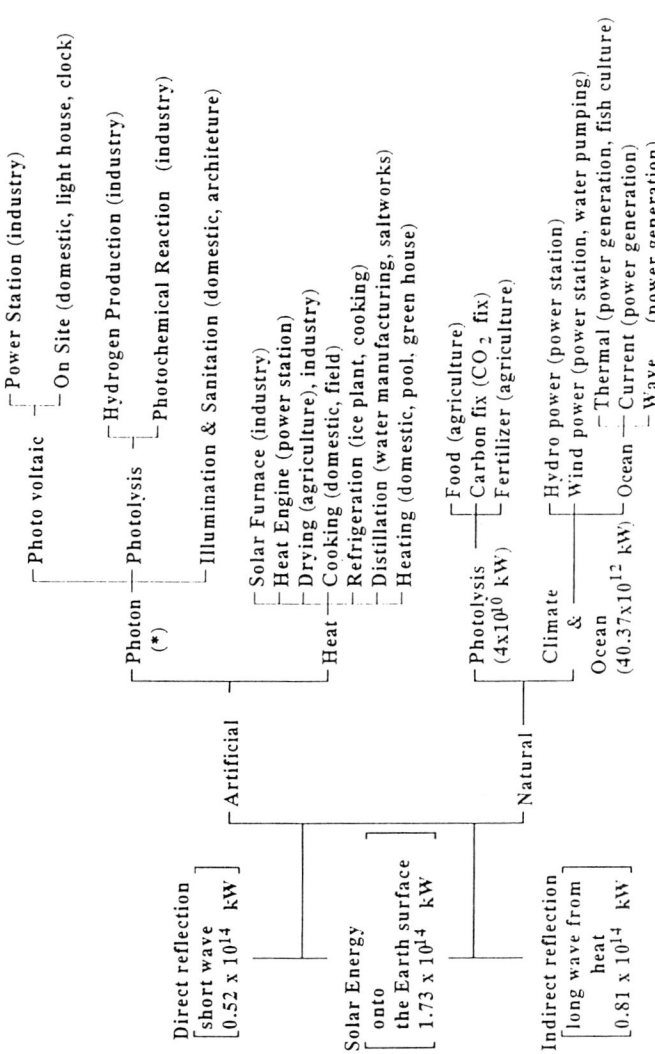

under illumination with 350 times of the collected solar beams at Boeing's High Tech-Center in 1990. The principles of tandem type as well as of common type solar cells will be described in a later chapter. Solar cells are commonly used in street lamps, lighthouses clocks, calculators, and so on. Recently, domestic power stations equipped with solar cells have been implemented in Europe, U.S.A., and Japan where preferential tax policies or financial aid systems are given by the government. At the end of 1994, the world capacity of working solar cells was 57.9×10^3 [kW]. The popularization of solar cells is a function of price. In Japan where the cost of power generation is the highest in the world the price of generating solar electric power is more than ten times that due to fire power plants. Solar cells can be said to be the standard-bearer of new energy technologies because it has a bright future to match these high-technologies. The urgent aim of development is to reduce the cost of power generating systems, including the solar cell itself.

Another promising and important technology for applying solar photon energy is the decomposition of water. This use is termed the "Solar - Hydrogen Energy System"[4,23] and utilizes the evolved hydrogen as a clean and powerful energy carrier in cooperation with other electric power systems. These new systems will be discussed in a later chapter.

The technical term "photolysis" includes not only water decomposition by photon energy but also any photochemical reaction used to obtain the desired products.

The traditional methods of illumination - and sterilization - using the sun's rays are being re-examined since the invention of optical fibers. These provide a pathway to lead solar beams almost anywhere and therefore makes them ideal for use in daily life.

Solar heat is the heat generated by the high temperature collection of solar energy. The value of the heat is shown by the exergy theory to be proportional to the square of the temperature difference. Therefore the heat generated using a Fresnel lens or concave mirror has the capacity for doing work. The solar beams incident on a Fresnel lens are focused at a point where the entropy of the system is greatly reduced. If the temperature of the focused place is 300°C and the ambient temperature is 27°C then the entropy of the focus is reduced by about half. We like to use this focusing phenomenon. However, the lens system subsequently diverges the solar beams after focusing, thus increasing the entropy. Searching for similar, entropy reducing natural phenomena is an important task in energy science.

The application fields listed in Fig. 1.7 are well-known and rather traditional but new technology is having an impact and will eventually put into practical use. Examples of such technologies include membrane, heat pipe, heat pump, new functional materials, and so on, which will be discussed later.

The photolysis effects in the natural world are numerous. Agriculture has a photolysis mechanism as its base. We should notice that fossil fuels are the result of the anabolism of green plants. Very few scientists[18] believe the theory of an inorganic source of petroleum. It should be stressed that green plants have a great capacity for eliminating atmospheric carbon dioxide and consequently they will have a great role in tomorrow's global environment.

Figure 1.7 shows that 23.3 % of the solar energy absorbed by the earth's surface atmosphere activates climatic and also ocean phenomena. These are rainfall, winds, storms, waves, ocean currents and ocean thermal phenomena. The utilization of these sources is described below.

(2) Hydro power, wind power, and ocean energy

Other energy converters of wave motions have been investigated; however, the period of **(a) Hydro power.** According to Lvovitch (Russia), the world's total annual rainfall is, on average, 108.4×10^{12} [t/Y], of which 12×10^{12} [t] are absorbed into the ground, 25.13×10^{12} [t] are carried away to the sea, and 71.27×10^{12} [t] are evaporated. If the quantity of rain water described above falls from a height of 1,000 [m] above the earth's surface, then a kinetic energy of $1,062.32 \times 10^{15}$ [kJ/Y] is imparted to the earth every year. Some of this huge energy is stored at dams, constructed in mountain valleys, confining the potential energy so that one can utilize it to generate electric power sufficient for their needs. Such a system is called the "dam system". There also exists another system called the "flow system" or "river system". The latter system utilizes the kinetic energy possessed by the flowing water of a river, and is constructed in the river. This system consists of the intake of river water, a precipitation area where muddy water is purified, and intake pipes leading to the water turbines. The water head is not so large as that of dam power systems.

A typical river system power station is the Bonneville power station existing on the Columbia river along the boundary between the states of Oregon and Washington in the U.S.A. This power station has a capacity of $ca.$ 1.06×10^6 [kW]. Beside the Bonneville station, there exist four more large river system power stations. The total sum of their outputs is over 8.4×10^6 [kW].

Dam system power stations can hardly find suitable sites in developed countries, while large reserves of river system power remain untapped in the underdeveloped countries.

The world's reserves of hydro power for both types of power generating system are estimated to be about 20×10^8 [kW], only about 10 to 15 % of which are developed while the remainder exist in developing countries such as China, India, South America, Siberia, and Canada. Most of them are of the river system type. In 1991, China proposed the construction of the Chang Kiang river dam that is 2 [km] long, with a 200 [m] height between the highest water level of the upper stream and the power station site. This dam can store 730×10^8 [t] water and has the capacity to generate 25×10^6 [kW] of power, the largest in the world. It is said that more than one million people had to move from the dam sites.

Another example of the social problem accompanying dam construction is that of Narmada river dam in India where more than one hundred thousand people were forced to move. Such a problem is, more or less, typical of any dam construction.

A new type of hydro power station is the pumping dam which has been invented as a storage method of electrical energy. Being different from oil-, coal-, and gas-fired plants, it is barely possible (actually impossible) to quickly alter the output of a nuclear reactor. Therefore the output from an atomic power station is usually constant even at midnight or on holidays. Thus it is necessary to construct a acility for storing electrical energy when nuclear power plants are constructed.

The mechanism of energy storage in a pumping dam is obvious. The surplus electrical energy drives motors to pump water up from a pond to a dam located a few hundred meters higher than the pond. Then, when electrical energy is needed, the water is released and hydro power drives turbines to generate the needed power. It is an advantage that the pumping motor can be used also as a generator. The system efficiency of the pumping dam is about 30 %. In 1990 Japan had a pumping station capacity of 5.63×10^6 [kW] while that of its other hydro power stations was 20.30×10^6 [kW].

(b) Wind energy. The windmill is one of the most traditional and oldest of human technologies. It is inferred that the invention of the windmill on which white sails are stretched occurred about 2,000 B.C. where it was utilized on the Aegean islands and on Crete. The spreading of windmills was most lively in the Netherlands, Denmark, and Northern Germany where there are low lands.

According to U.A. Coty's (Lockheed - California) report of 1976, the wind power resource of the U.S.A. is about 17.2×10^8 [kW]. However, the estimation of

Table 1.7. Distribution of windmills in the countries of IEA (1992)

Country	Capacity (10^3 kW)	
	working	under construction
U.S.A	1,600	260
Denmark	410	100
Germany	100	0
Netherland	80	10
England	10	131
Sweden	6	4
Canada	5	10
Italy	2	20
Japan	2	2
Norway	0	2
Total	2,200	572

Energy and Its Resource 37

the resource can vary considerably with location so it is not fair to say that the ratio of land area reflects the quantity of the resource.

In 1992 the world output from installed windmills was about 2.22×10^6 [kW]. The distribution of these windmills is shown in Table 1.7. However, at the end of 1993, wind farms built in California have 19,000 windmills generating 190×10^6 [kW].

The many kinds of windmills are shown in Table 1.8. These include propeller and other axial-flow turbines, as well as radial systems mounted on vertical axes. Most modern windmills have two or three blades which have the highest efficiency under most climatic conditions. In order to generate alternating current

Table 1.8. Basic classification of windmills

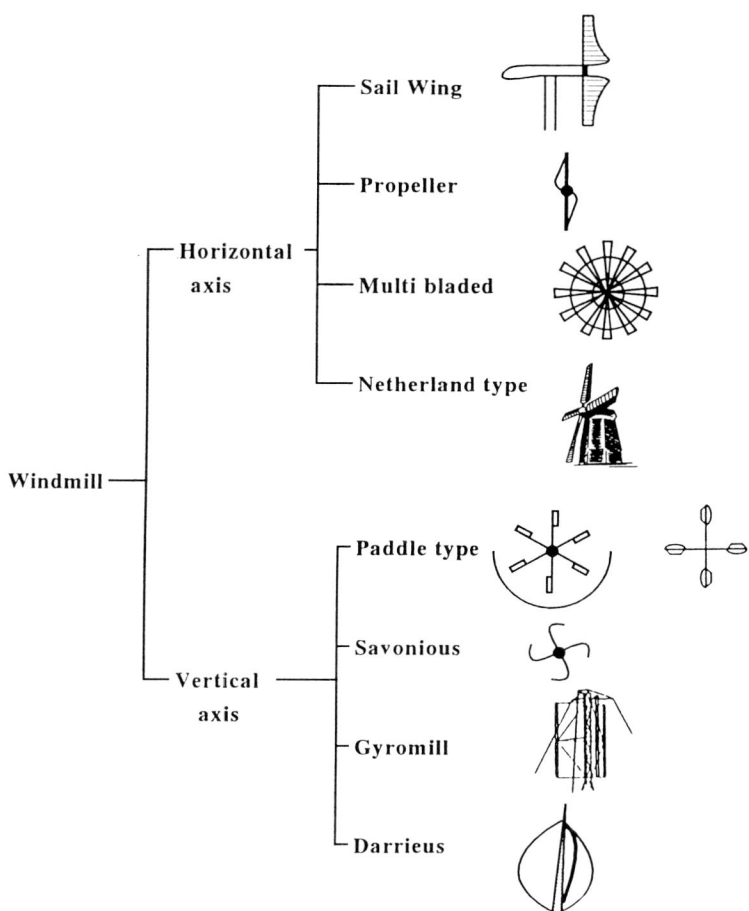

38 Energy Technology

electric power with a constant cycle, the windmill must be operated at a constant angular velocity over a wide range of wind speeds. In addition, other contrivances are necessary, for example, an energy storage battery system is necessary when no wind comes, automatic stopping systems are needed when too strong a wind comes, and so on.

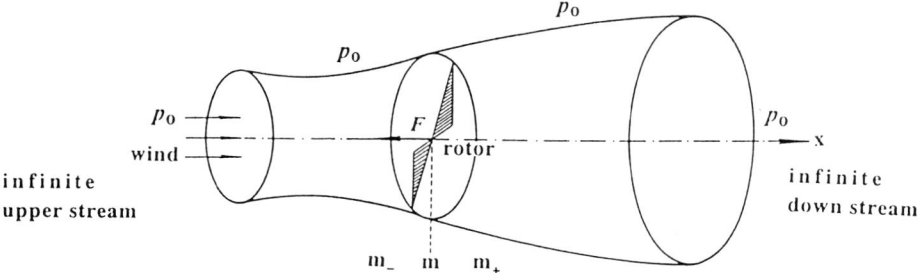

Fig. 1.8. The principle of the windmill.

We shall now present a fundamental theory of windmills that can be understood using elementary fluid dynamics. First of all, we will define technical terms that are necessary in the derivation. Velocity is a vector at every point in the fluid, accordingly we can define a curve whose tangent at each point is a velocity vector. Such a curve is called a stream line. A fixed number of stream lines is called a stream tube where the cross section is subject to change but the number of lines is not. If the fluid is incompressible then we have

$$vS = \text{const.}, \tag{1.72}$$

where v is the fluid velocity and S is the cross section of the stream tube.

The kinetic energy of the wind per unit time (wind power) is

$$W_l = \frac{1}{2} (\rho v S) v^2$$

$$= \frac{1}{2} (\rho S) v^3 \tag{1.73}$$

where ρ is the air density (=1.2 [kg/m³] at 20°C and atmospheric pressure). Equation (1.73) shows that the wattage of a windmill is (i) proportional to the third power of the velocity, and (ii) proportional to the square of the rotor radius.

Next, we shall show that the maximum efficiency of a windmill is

$$\eta_m = 16/27.$$

In Fig. 1.8, the notations m, m_, m+, and p_0 indicate the rotor plane, just before and after the rotor plane, and the atmospheric pressure, respectively. The quantity of flowing air is given by

$$Q = S_1 v_1 = S_m v_m = S_2 v_2, \tag{1.74}$$

where subscripts 1, m, and 2 indicate the variables at an infinite upstream location, the rotor plane, and at an infinite downstream location, respectively. The wind outside the stream tube is uniform while the velocity and pressure inside the stream tube changes. The spinning rotor imparts a force F to the flow tube, the magnitude of which is

$$F = Q(v_1 - v_2)$$
$$= S_m(p_{m-} - p_{m+}). \tag{1.75}$$

From Eqs. (1.74) and (1.75), we have

$$p_{m-} - p_{m+} = r v_m (v_1 - v_2) \tag{1.76}$$

According to Bernoulli's theorem (p.13), we have

$$p_0 + \frac{1}{2}\rho v_1^2 = p_{m-} + \frac{1}{2}\rho v_m^2,$$

$$p_0 + \frac{1}{2}\rho v_2^2 = p_{m+} + \frac{1}{2}\rho v_m^2, \tag{1.77}$$

from these equations, we have

$$p_{m-} - p_{m+} = \frac{1}{2}\rho(v_1^2 - v_2^2) \tag{1.78}$$

from Eqs. (1.76) and (1.78), we get

$$v_m = \frac{1}{2}(v_1 + v_2). \tag{1.79}$$

The wattage needed to drive the rotor is given by

$$W = (p_{m-} - p_{m+}) v_m S_m$$
$$= \rho v_m^2 (v_1 - v_2) S_m. \tag{1.80}$$

If the wind velocity v_1 is reduced by an amount a by driving the rotors, we have

$$v_m = (1-a)v_1 \tag{1.81}$$

and from Eq. (1.79), we get

$$v_2 = (1-2a)v_1. \tag{1.82}$$

This relationship means that the velocity at an infinite downstream location is reduced by twice the reduced rate just after the rotor's plane. Substituting Eqs. (1.81) and (1.82) into Eq.(1.80), we have

$$W = 2\rho a (1-a)^2 v_1^3 S_m. \tag{1.83}$$

This is the fundamental formula governing windmill performance. The maximum wattage is derived from this equation by the following procedure.

The condition of maximum power is

$$dW/da = \text{const.}(1-3a)(1-a)$$
$$= 0. \tag{1.84}$$

where a must not be 1 (if so then wind disappears at the rotor's plane), therefore we have, for the most efficient parameter a,

$$a^* = 1/3. \tag{1.85}$$

Substituting the value $a^* = 1/3$ into Eq.(1.83), we get

$$C_p = W/W_l$$
$$= 4a^*(1-a^*)^2 = 16/27, \tag{1.86}$$

which is called the "Betz's Law" and shows that the maximum efficiency cannot exceed the value $16/27 = 0.593$. The ratio C_p is called the induction coefficient.

Windmill power generation has many problems such as spoilage by salty wind and attachment of ice and snow on the blades. Technologies for manufacturing blades using new materials made of glass fibre, carbon fibre, and so on, which avoid such damage, are important. Computer applied control is another necessity.

The social problems of windmills are, for example, that the electromagnetic waves for TV, radio and other communications are disturbed by the spinning rotors; counterplans are necessary.

(c) Ocean energy. Wave energy, ocean current energy, and ocean thermal energy belong to this realm. There is no practically applied example in this field yet, although the energy resources are numerous as inferred from Fig.1.7.

(i) *Wave energy.* The energy density of ocean waves ranges widely, from 10^{-3} to 10^3 [kW/m], where the energy unit shows the energy per 1 [m] of coastline. The average energy of a wave with amplitude A is

$$\overline{W} = \frac{1}{2}\rho g A^2 \qquad (1.87)$$

$$= \frac{1}{8}\rho g H^2, \qquad (1.88)$$

where $H = 2A$ is the height of the wave.

The total amount of wave energy in the world has been studied by many authors. Their results suggest that the total resources are of the order of 10^9 [kW]. This numerical value can be confirmed by inserting the mean value of H (= 1.5 [m]) into Eq. (1.88) and integrating over all the coastlines in the world. This integral gives 2.7×10^9 [kW].

There are a several kinds of effective energy conversion systems, from wave energy to electrical energy. Among them, the most popular method is to activate air turbines. If a pipe equipped with an air turbine at the upper end is fixed in the sea water near a coast, the wave motion periodically raises and lowers the water level in the tube. Accordingly the air is either pressed out or drawn in as the water level rises and falls. Thus the turbines are activated to generate electric power (Fig. 1.9).

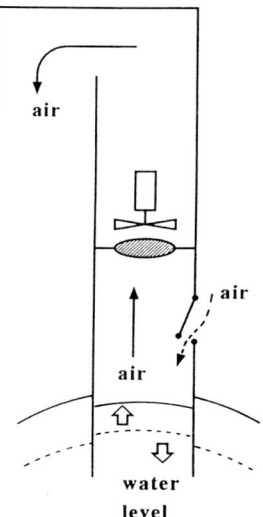

Fig. 1.9. Turbine activation by wave energy.

the sea waves is not always constant so that the generated electrical energy can hardly have a constant cycle. The energy density of a sea wave is not high, on average, while a conversion facility requires a big investment. Hence, it is not easy to develop a wave power plant that is practical.

(ii) *Ocean current energy.* The solar energy falling upon the tropical zone is so intense and plentiful that the sea water is warmed and becomes a warm current that flows towards the southern and northern poles. The warm current is cooled by ice in the Antarctic and Arctic seas and flows back to the tropical sea. This is the heat cycle of tropical solar energy. Both currents flow along nearly the same path every year. Therefore we may extract kinetic energy using water turbines that are fixed in the sea (*e.g.*, moored between islands). It is probably not feasible to construct power plants utilizing such ocean currents as the investment would be too great.

(iii) *Ocean thermal energy.* As described above, the two kinds of ocean currents exist, one being warm and the other cold. Dr. A. d'Arsonval, a French physicist, first proposed an energy conversion system using the temperature gradient between them and his student, G. Claude, made a preliminary experiment as early as in 1881.

The principle is simply described as follows. A heat medium liquid such as CFC (Chlorofluorocarbon) or ammonia is cooled to the liquid state by deep sea water at a temperature of 5 - 10°C. The liquid is then pumped up from the deep sea and vaporized by the warm surface current where the temperature is 20 - 25°C. The resulting vapor pressure is used to drive a gas turbine. This system is called OTEC (Ocean Thermal Energy Conversion).

The Carnot efficiency of a system is written as

$$\eta_c = (T_1 - T_2) / T_1 \tag{1.89}$$

where T_1 and T_2 are its higher and lower reservoir temperatures. This is highest whenever heat energy is converted to any other kind of energy. In an OTEC, the highest possible efficiency is only $\eta_c = 20/300 = 0.07$. In addition, the necessary energy to pump up the deep sea water will consume more than half of what is generated. These and other losses are of course unavoidable, hence the generated net electric power will only be about 1 % at most. Some preliminary experiments were carried out in the U.S.A. and Japan in the 1980s.

The electric power generated in this system can be used to electrolyze sea water to generate hydrogen which can then be easily transported, instead of using direct electrical transmission which is difficult in mid-ocean.

Despite its low energy efficiency, this ocean system has another merit of pumping up nutritious deep sea water to support the on-site fish culture as mentioned before.

(3) Geothermal energy

Geothermal resources are abundant as shown in Table 1.6. Some representative geothermal power plants are listed in Table 1.9. Since the oil crises of 1973 and the 1980s, several more power plants have been developed; we currently have about 4×10^6 [kW] as the total world capacity of the approximately 100 electric power plants' output.

To enhance efficiency, binary cyclic power generation has been developed. This system is composed of two generation processes, one of which uses a gas turbine driven by flon or ammonia and the other a water vapor utilization turbine.

Geothermal systems usually need both underground water and magma heat. However, there are few places where both coexist. Therefore some new systems have been investigated, for example, magma or heated rock systems which utilize artificially injected water to obtain hot steam.

High grade technologies will be applied first to identify suitable sites. LANDSAT, an artificial satellite, searches using infrared rays. Shallow resources (100 - 200 [m], 100 -250 °C) are likely to be found in this way, deeper resources being more difficult to identify. Magnetic research using, for example, Josephson's device which is very sensitive to magnetism, and other sophisticate instruments can be used. Another application of future technology is to effectively remove precipitate and chemical material attached to the vapor pipe lines.

Suitable sites for geothermal power plants are along volcanic zones where magma exists near to the earth's surface. These mountainous regions have fine landscapes and are often designated as National Parks, so that development is nearly impossible. Most of the suitable sites belong to such areas.

The public pollution due to geothermal utilization results from the exhausted water containing sulfur hydrides that damages wood and glasses. Recently, the exhausted water has been recovered and poured into the original underground holes. This system has another merit in that the underground water is not depletable.

Table 1.9. Representative geothermal power plants (1980)

Field	Country	Capacity 10^3 kW
Cerro Prieto	Mexico	630
The Geysers	U.S.A.	1,700
Kakkonda	Japan	55
Larderello	Italy	410
Namfjall	Iceland	3
Pauzhetsk	Russia	5.7
Wairakei	New Zealand	202

(4) Lunar tide energy

The periodical movements of the ocean water by the lunar attractive force are known as tide phenomena. The differences between the sea levels of ebb tide and high tide vary from coast to coast. The largest head observed thus far is about 16 [m] at Fandy bay on the east-southern coast of Canada. The only realized practical power plant is the Rance power station in France where the largest head is 13.5 [m] and the high tide visits every 6 hours and 12 minutes.

The late President C. de Gaulle (1890 - 1970) of France made up his mind to construct a tidal power station at the mouth of the river Rance in 1961. The station, finished in 1966, is composed of 24 water turbines, each of which has a capacity of 10^4 [kW]; it generates 5.48×10^8 [kWh] in a year. The bank between the gulf of St. Malo and the artificial dam has a length of 300 [m], height of 12.3[m], and a width of 20 [m]. Under this bank the 24 turbines are equipped to generate electric power when the high tide comes in and also when the water leaves the bay.

China has also constructed several tidal power stations on the coast facing the Po Hai bay. Their output is estimated to be 6,000 [kW].

There are many suitable sites where the high tide occurs, especially in Canada, Northern Europe, and Northern Russia. However, the investment cost is so expensive that no government has undertaken the efforts except for President de Gaulle.

Chapter 2
Energy Conversion

Steam-power station

Chap. 2. Energy Conversion

Classifications of energy were considered in Section 1-2 of the previous chapter, where we list the five kinds of energy. Energy conversion among them have been studied in both science and technology. The innovation of conversion technology has exerted great influence upon human civilization. Representative examples are the Industrial Revolution due to the invention of steam engine and the release of nuclear energy. Alongside such big and striking events small and inconspicuous improvements are continually made.

The classified energies have their own characteristics through which the energy conversions are carried out. A systematic classification is due to the late T. Takahashi (Tokyo University), who was a prominent leader of energy science in Japan. We shall describe a plain but rigorous theory according to Takahashi but, before introducing this theory, we will make some basic comments on energy conversion.

There exist two kinds of energy conversions. The first is called direct energy conversion. The initial type of energy is directly converted to the final one without any mediated energy; for example, there is thermoelectric conversion (from heat to electricity), fuel cell (from chemical energy to electricity), solar cell (from photon energy to electrical energy), thermoionic conversion (from heat to electricity), magneto hydrodynamic conversion (from heat to electricity) and so on. In these conversions, no third kind of energy is mediated.

Direct energy conversion has, in principle, a relatively high efficiency and produces less entropy and is therefore recommended for future technology. We shall discuss this in this chapter.

The second kind of energy conversion is called indirect energy conversion. Mechanical energy and electrical energy can be converted with mutually high efficiency, because this energy conversion does not always accompany entropy production. Therefore, in order to get electricity, one may convert the initial energy to the mechanical energy and then obtain electricity from a power generator. In typical fire powered plants, the chemical energy of the fossil fuel is changed to heat, then this heat makes water vapor that activates a turbine to generate electric power. That is, the following process:

initial chemical energy $\to$ heat $\to$ mechanical $\to$ final electrical energy,

is realized. On the other hand, a fuel cell has the process:

initial chemical energy $\to$ final electrical energy.

Entropy production is appreciable in the conversion of chemical energy to heat and from heat to mechanical energy, and therefore the fuel cell is preferred.

2-1. Matrix of Energy Conversions

As has often been mentioned, there are five kinds of energies so that at least twenty five kinds of energy conversions exist. These are arranged in Table 2.1 as a matrix. The elements of the matrix are examples of conversion phenomena or technologies. Most of them are familiar to the readers. However, some of them are curious and their explanations are presented later.

Some of the elements represent direct energy conversions, but others do not.

(a) Tribo-luminescence. This effect has an alternative name of tribo thermal luminescence which means that the heat caused by the mechanical energy of friction gives rise to luminescence. However, the impulse or the friction may directly cause this effect although the mechanism is not clear yet.

Table 2.1. Matrix of energy conversions

After \ Before	Mechanical Energy	Electrical Energy	Chemical Energy	Photon Energy	Heat Energy
Mechanical Energy	· Torque convertor · Fly wheel	· Generator · Piezo electricity · M.H.D.(*)	· Mechano chemical effect · Metalhydride	· Tribo-luminessence	· Friction · Collision · Metalhydride
Electrical Energy	· Motor · Linear Motor · Magneto-electro striction	· Transformer · Invertor · Microwave transmission	· Electrolysis · Electro-chemical reaction	· Electro-luminessence · Laser	· Joulian heat · Peltier heat · Microwave absorption
Chemical Energy	· Mechano chemical effect · Explosion · Super expansion · Metalhydride	· Primary battery · Secondary battery · Density gradient power generation	· Chemical reaction	· Chemical luminessence · Chemical laser	· Combustion · Dillution heat · Metalhydride
Photon Energy	· Light pressure · Photon rocket	· Photoelectric effect · Solar cell	· Photo chemical reaction · Photo electrode reaction	· Maser · Fluoressence · Phosphorescence	· Absorption · Norbornadiene
Heat Energy	· Heat engine · Convection · Shape memory · Metalhydride	· Seebeck effect · A.M.T.C(**) · Thermo electron emission · Thermo dielectric conversion	· Thermo-chemical reaction · Distillation · Metalhydride	· Radiation	· Heat pump

(b) Mechano chemical effect. If the application of an impulse, or pressure to a chemical system initates its reaction, then such an effect is called the mechano chemical effect. The reverse effect is also possible.

(c) Super expansion. Wood expands by absorbing water. The water molecules enter the spaces in the network structure of the wood's molecules. Agar and jelly also have large molecules and a network structure that can appreciably expand by absorbing water. They are called gels. If polyacrylamide (a polymer of C_3H_5NO) is submerged in alcohol or acetone, then it assumes a jelly-like state and under some conditions achieves an extraordinary expansion. For example, a polyacrylamide specimen submerged in an aqueous solution of acetone makes an extraordinary contraction to below 1/350 of its original volume when the density of acetone increases to 42 %. Then, if the temperature rises from water temperature to 22°C, while maintaining a 42 % acetone density, it rapidly returns to the original volume. This extraordinary expanding phenomena can also be controlled by applying an electrical voltage instead of temperature.

(d) Primary- and secondary-battery. Batteries are indispensable in the age of modern electronics. An innovative hydrogen battery will be introduced in Chapter 4, but we shall mention only the differences between them here.

Primary batteries are the most widely used. Common examples of this type are zinc-carbon and alkaline-manganese dry cells. These are not designed to be rechargeable.

A secondary battery can be recharged. Common examples of this type are lead-acid and nickel-cadmium batteries, which are widely used in conjunction with internal combustion engines.

The fuel cell belongs to this type and will be discussed in Chapter 4.

(e) Norbor-nadine (C_7H_8). When irradiated with light, the norbor-nadine isomerizes to quadricylene which has the same molecular formula but a different molecular structure. The structure of norbor-nadine is represented:

This structure is deformed with strain by absorbing light to form the quadricylene. However, if a catalyzer is present, the deformation is recovered giving off the potential energy. A new material was synthesized in 1982 by professor Z. Yoshida (Kyoto University) by adding methyl- and cyanogen-groups. This new material can store solar heat which is evolved by a catalyzer when needed. It is a very efficient way of storing heat; 1 [*l*] of the material can store about 59 [kcal] of heat.

(f) Shape-memory alloy. This group of alloys was first found by a group at the Naval Ordnance Laboratory (U.S.A.) in 1963. It was a TiNi alloy and is called Nitinor ("nor" is the abbreviation of the said laboratory). In addition, Cu-Zn-Al, Fe-Zn-Al, Fe-Mn-Si, and other alloys are known. These alloys memorize their shapes above or below a critical temperature so that the original shape can be repeatedly reproduced from any shape deformed below or above the critical temperature. Applications of these alloys are multifarious. For example, we shall show a **heat engine**. Consider two kinds of shape memory alloys whose critical temperatures are different from each other. One of them is higher than and another is lower than room temperature, respectively. If these two kinds of alloy are connected to a rotary disk as shown in Fig. 2.1, then by heating the system the alloy-bar A will recover its original length pushing the disk from right to the left (in the direction of the arrow), while alloy-bar B is pressed and deformed. Next, if the system is cooled down, then alloy-bar B will recover its original length pushing the disk from left to right. Thus the disk makes a forward and return motion.

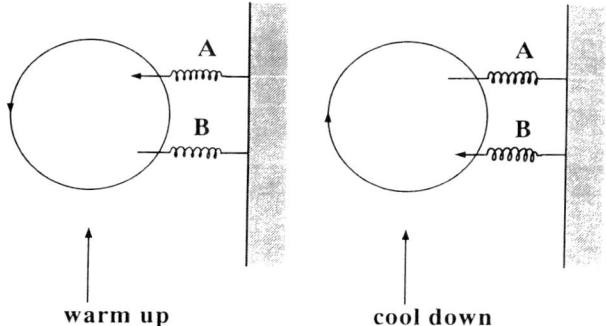

Fig. 2.1. An application of the shape memory alloys to a heat engine.

Besides the energy conversion described above, we have many other applications which are important in modern technologies. We shall name them according to the matrix in Table 2.1 and, all of them are explained in detail on the pages indicated in the attached bracket.
(i) Fly wheel (p.133), (ii) M.H.D. (magneto hydro dynamics, p.74), (iii) Metal hydride (p.138, 191), (iv) Electrolysis (p.149), (v) Solar cell (p.83), (vi) Photo electrode (p.176), (vii) AMTEC (Alkali Metal Thermoelectric Conversion, p.188), (viii) Thermo dielectric conversion (p.54), (ix) Density gradient power generation (p.183)

Some more devices are introduced in this book.

2-2. Quasi Static Process

A coordinate-system with an intensive variable (η) and an extensive variable (ξ) as the vertical and horizontal axes, respectively, is usually taken to indicate the physical state at a point in the system. Therefore, a path such as from P_1 to P_2 shows the process of changing from the state (ξ_1, η_1) to the state (ξ_2, η_2). If the starting state P_1 is merged with the final state P_2, then the process is called a reversible process, if not it is the irreversible process.

The concept of a quasi static process is an ideal approximation of a thermodynamic change. Every elemental change of the thermodynamical state of a system is so quiet and small enough that it is reversible. Accordingly, the total change of a system composed of an infinite number of these elemental changes is also considered reversible. We call such a system a quasi-static process. It is a method of approximation. Practical heat engines are studied on the basis of quasi static processes.

(1) Carnot's cycle

The technical term "cycle" indicates a process shown by a closed path in the ξ - η system plane as in Fig. 2.2. We shall consider a simple cycle for one mole of an ideal gas (this material is called the working material) as shown in Fig. 2.3, where four changes (processes) are combined. These are (i) isothermal expansion at higher temperature T_2 from state $A(p_1, V_1)$ to state $B(p_2, V_2)$, (ii) adiabatic

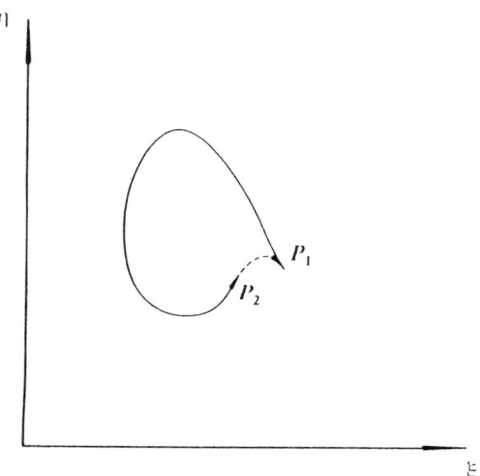

Fig. 2.2. The coordinate system with an intensive variable(η) and n extensive variable(ξ) as the horizontal and vertical axes, respectively. A point and a path in this space show a thermodynamical state and thermodynamical process of a physical system, respectively.

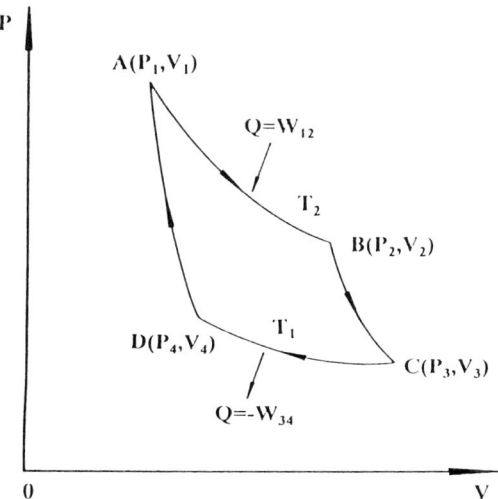

Fig. 2.3. Carnot's cycle.

expansion from $B(p_2, V_2)$ to the $C(p_3, V_3)$ at lower temperature T_1, (iii) isothermal compression from the state $C(p_3, V_3)$ to the state $D(p_4, V_4)$, and (iv) adiabatic compression from the state $D(p_4, V_4)$ to the original state $A(p_1, V_1)$.

The relationships between the states are as follows.

$$\left.\begin{array}{c} p_1 V_1 = p_2 V_2 = RT \\ \\ p_3 V_3 = p_4 V_4 = RT \end{array}\right\} \quad (2.1)$$

$$\left.\begin{array}{c} p_2 V_2^\gamma = p_3 V_3^\gamma \\ \\ p_4 V_4^\gamma = p_1 V_1^\gamma \end{array}\right\} \quad (2.2)$$

where T_1 and T_2 are the higher and lower temperatures, respectively, and R (= 8.31 [J/mol·K]) is the universal gas constant, and γ (= C_p / C_v) is the ratio of specific heats.

From Eqs.(2.1) and (2.2), we have

$$V_1/V_2 = V_3/V_4 \quad (2.3)$$

The input and output of heat and mechanical energy for each change is calculated as follows.

(i) *The input heat Q_2 is equal to the work W_{12} done in the process from A to B:*

$$Q_2 = W_{12} = \int_{V_1}^{V_2} p dV$$

$$= RT_2 \ln(V_2/V_1), \qquad (2.4)$$

(ii) *The output heat Q_1 is equal to the work W_{34} done by the external system in the process from C to D:*

$$Q_1 = -W_{14} = -\int_{V_3}^{V_4} p dV$$

$$= RT_1 \ln(V_3/V_4). \qquad (2.5)$$

(iii) *The work, W_{23}, from B to C is obtained when equation (2.2) is applied*

$$W_{23} = \int_{V_2}^{V_3} p dV$$

$$= [R/(\gamma - 1)](T_2 - T_1), \qquad (2.6)$$

(iv) *It is easily proved that W_{23} is equal to the work W_{41} from D to A,*

$$W_{41} = [R/(\gamma - 1)](T_1 - T_2). \qquad (2.7)$$

Therefore the total work of the ideal gas is given by

$$W = W_{12} + W_{23} + W_{34} + W_{41}$$

$$= Q_2 - Q_1$$

$$= R[T_2 \ln(V_2/V_1) - T_1 \ln(V_3/V_4)]$$

$$= R(T_2 - T_1)\ln(V_2/V_1), \qquad (2.8)$$

when Eq.(2.8) is combined with Eq.(2.4) we obtain the efficiency of Carnot cycles:

$$\eta_c = W/Q_2 = (T_2 - T_1)/T_2. \qquad (2.9)$$

This formula is called the Carnot efficiency and represents the maximum possible value of all heat engines operating between temperatures T_1 and T_2.

(2) Electric tray

An electric tray is a traditional experimental apparatus to study the conversion between mechanical and electrical energies. Two parallel plates made of electrical conductors are connected through a switch to two batteries whose voltages are V_1 and V_2, respectively (Fig. 2.4). The separation x between the plates can be varied so that the capacitance C of the condenser also varies to change the potential V between them. The fundamental relationship is

$$V = q/C$$

$$= xq/\varepsilon S, \qquad (2.10)$$

where q, ε and S represent the charge, dielectric constant of air, and the surface area of the plate.

We shall consider a cycle, referring to the q-V coordinate systems shown in Fig 2.5.

(i) *We start from $B(q_2,V_1)$ to $C(q_2,V_2)$ by increasing x from x_1 to x_2 keeping q ($= q_2$) constant (the switch is put to the off state).* The voltage V increases from V_1 to V_2, and mechanical energy is converted to the electrical potential energy $q_2(V_2 - V_1)$.

(ii) *Next, we proceed to the process from $C(q_2,V_2)$ to $D(q_1,V_2)$.* The switch S is connected to battery V_2 and x is increased from x_2 to x_3. This operation decreases the capacitance of C and the charge decreases from q_2 to q_1 because the voltage is maintained at V_2. The work done by the external system is equal to the electrical energy $(q_2 - q_1)/V_2$.

(iii) *The process from $D(q_1,V_2)$ to $A(q_1,V_1)$ is now considered.* The switch S is in the open state. The interval x decreases from x_3 to x_4. The charge remains constant

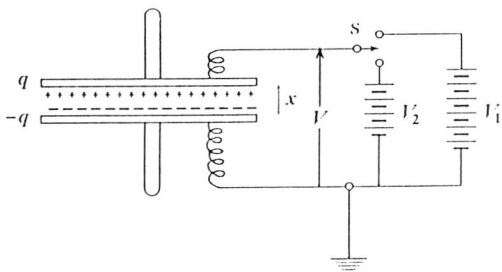

Fig. 2.4. An electric tray. The interval between parallel plates can be variable (x changes). They are charged from higher voltage batteries to get charge q.

while C increases and the voltage decreases from V_2 to V_1. The magnitude of the work done to the external system is equal to $q_1(V_2 - V_1)$.

(iv) *The final process is from $A(q_1, V_1)$ to $B(q_2, V_1)$.* The switch is connected to the battery V_1 and the charge $(q_2 - q_1)$ flows into the plates. Work is done to the external system with a magnitude of $(q_2 - q_1)V_1$.

A concise description of the above cycle $A \rightarrow B \rightarrow C \rightarrow D \rightarrow A$ is that the charge $(q_2 - q_1)$ is given to the battery V_2 by picking up charge from a battery with a lower voltage V_1. That is to say, the charge $(q_2 - q_1)$ is carried up from V_1 to V_2. The work done by this tray system is $(q_2 - q_1)(V_2 - V_1)$.

The conversion efficiency is

$$\eta_T = \text{output in mechanical/input in electrical}$$

$$= (q_2 - q_1)(V_2 - V_1) / V_2 (q_2 - q_1) = (V_2 - V_1)/V_2. \quad (2.11)$$

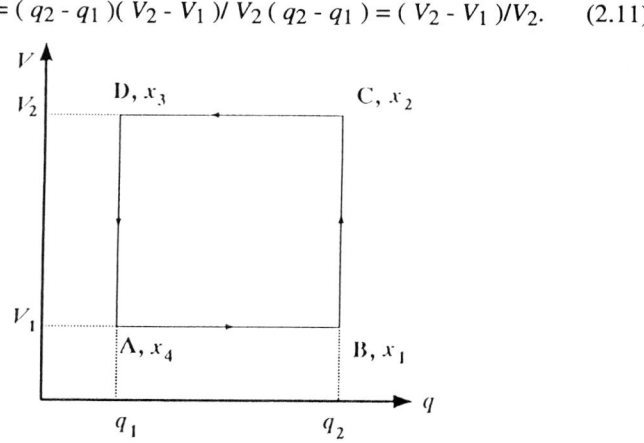

Fig. 2.5. Cycle of electric tray in the $q - V$ space.

(3) Thermo-dielectric conversion[8]

The electric tray is used to convert between electrical and mechanical energies. To vary the capacitance of the condenser, a mechanical displacement is introduced; on the other hand, if thermal energy is used to vary the capacitance, then a similar conversion system is possible.

An electrical thermo-dielectric conversion circuit is shown in Fig.2.6. The same relationship as in Eq.(2.10) is held between charge q_1, capacitance C, surface area S, distance between plates x, and dielectric constant of the medium between the plates ε.

Ferroelectrics such as $BaTiO_3$, $PbNbO_3$, $LiNbO_3$, $LiTaO_3$, $SbSI$, and so on are used as the dielectric medium. Their dielectric constants are expressed by

$$\varepsilon = \varepsilon_c + C/(T - T_c), \quad (2.12)$$

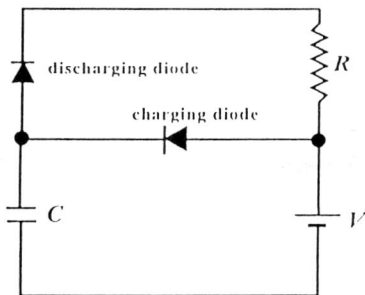

Fig. 2.6. Electrical circuit of thermo-dielectric conversion.

where C is the Curie - Weiss constant and T_c is the Curie temperature. Taking two temperatures T_1 and T_2 ($>T_1$), the following relationship is obtained, from Eqs. (2.10) and (1.15),

$$V_2/V_1 = \varepsilon_1/\varepsilon_2$$

$$= W_2/W_1 \qquad (2.13)$$

where $V_{1,2}$ and $W_{1,2}$ are the voltages and stored electrostatic energies respectively.

Equation (2.13) indicates that heat energy makes the dielectric constant decrease so that the voltage between two plates as well as the stored energy increase.

We shall consider the following cycle (Fig. 2.7).

(i) *A charging process from $A(0,0)$ to $B(q,V_1)$ at temperature T_1.* This is an isothermal charging process. The stored energy in this process W_1 is

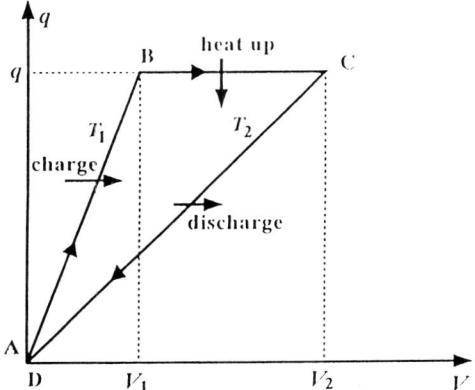

Fig. 2.7. Cycle of thermo dielectric conversion.

56   Energy Technology

$$W_1 = \int_0^q V dq$$

$$= Sx \int_0^{D_1} \frac{1}{\varepsilon(T_1)} D dD, \qquad (2.14)$$

where $D = \varepsilon E = \varepsilon V/x$, $q = SD$.

(ii) *A heating process from $B(q, V_1)$ to $C(q, V_2)$, where the temperature rises from T_1 to T_2.* The heat energy is converted to an electrical energy of $q(V_2 - V_1)$.

(iii) *A discharging process from $C(q, V_2)$ to $D(0, 0)$.* The condenser with charge q and voltage V_2 is discharged at the temperature T_2 and assumes a state with zero charge and voltage, but temperature is still T_2. This is an isothermal change and returns the charge given by process (i) to the battery. If any surplus electrical energy remains, then it can do work at R.

The work done by the condenser in this process is given by

$$W_2 = Sx \int_{D_1}^0 \frac{1}{\varepsilon(T_2)} D dD. \qquad (2.15)$$

(iv) *The cooling process from $D(T_3)$ to $A(T_1)$.*

Now, we consider the efficiency of this cycle. The electrical energy per unit volume, converted from heat energy by this cycle is

$$W = W_2 - W_1 = \int_0^{D_1} \left\{ \frac{1}{\varepsilon(T_2)} - \frac{1}{\varepsilon(T_1)} \right\} D dD.$$

$$= (D_1^2/2\alpha)(T_2 - T_1), \qquad (2.16)$$

where ε is approximated by $\varepsilon = \alpha/T$.

On the other hand, the input heat q, delivered to the two parts of the condenser, can be expressed by

$$q = \int_{T_1}^{T_2} C_v(D, T) dT + T_2 \int_0^{D_1} \frac{\partial}{\partial T} \left\{ \frac{1}{\varepsilon(D,T)} \right\} D dD$$

$$= C_v(T_2 - T_1) + T_2(D_1^2/2\alpha), \qquad (2.17)$$

where C_v is the heat capacity of the condenser. These integrals were carried out by assuming that C_v and ε are independent of T and D, respectively. The efficiency of this cycle is given by

$$e = W/q = [(D_1^2/2\alpha)(T_2 - T_1)]/[(D_1^2/2\alpha)T_2 + C_v(T_2 - T_1)]$$

$$\rightarrow (T_2 - T_1)/T_2. \qquad (2.18)$$

assuming that the heat capacity can be neglected.

Some experiments using $BaTiO_3$ were undertaken[44] at Osaka University and the result is that an efficiency of $e = 0.8$ % was observed under the conditions of $T_1 = 125°C$, $T_2 = 165°C$, and $D_1 = 0.2$ [C/m^2]. The electrical energy given by this conversion is not necessarily large but voltages as high as 1×10^6 [V] are readily obtained.

We have studied three examples of quasi-static energy conversion and obtained the efficiency;

$$e = (\eta_2 - \eta_1)/\eta_2, \qquad (2.19)$$

where η represents the intensive variable. The η_1 and η_2 represent the minimum and maximum magnitude used in the cycle respectively. If the efficiency of a quasi static cycle cannot be expressed by Eq. (2.19), then some appreciable energy loss, such as the heat needed to heat up the condenser in a thermo dielectric conversion, exists.

We may conclude, in general, that *the maximum efficiency in any quasi static energy conversion can be expressed by Eq (2.19)*.

In addition to the three energy conversions cited, some other phenomena are known. These are shown in Table 2.2. Most of them are well-known but some of them are not as familiar, for example, Wiedemann's effect. It is explained in Fig. 2.8.

The efficiency of the conversions listed in Table 2.2 has not always been calculated, but most of them are expected to follow the formula given by Eq.(2.19).

Table 2.2. Examples of quasi static conversions

	Thermal Energy	Electrical Energy	Magnetic Energy	Chemical Energy
Mechanical Energy	Heat expansion	Piezo-electricity	Wiedemann's energy effect	Osmotic pressure
Thermal Energy		Pyro-electricity	Thermo-magnetic effects	Thermo-chemical effects
Electrical Energy			Electro magnetic induction	Electrode potential

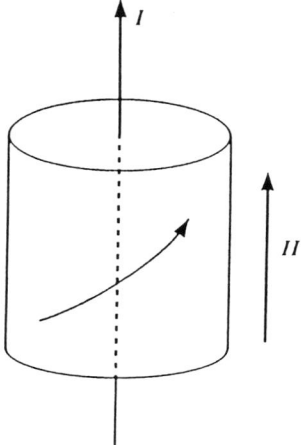

Fig. 2.8. Wiedemann's effect. G. Wiedemann (1826-1899) found that a metallic cylinder placed parallel to a magnetic field H is distorted when an electric current I flows in a wire along the axis of the cylinder. This is due to a screw magnetic field around there.

(4) Conditions for the possibility of quasi-static conversions

The efficiency of the quasi static conversion expressed by Eq.(2.19) shows that more input energy flows into the working substance than energy output, that is to say, some part of the input energy is not converted. This is called the "unavailable energy" which is different from the dispersed and entropy-increasing energy produced during the energy conversion. It is important to distinguish both, because the unavailable energy can be utilized repeatedly by other processes, while the dispersed energy can hardly be used.

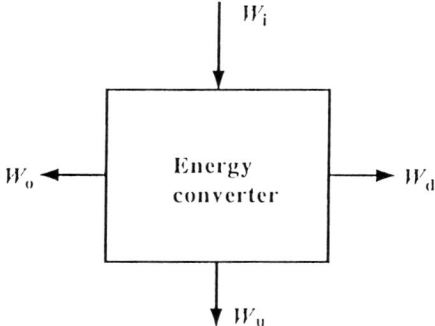

Fig. 2.9. Input, output, unavailable, and dispersed energies. These energies are indicated by subscripts i, o, u, and d, respectively.

One of the most prominent features of quasi static conversion is the remaining unavailable energy. The balance of the input energy W_i, the output (aimed) energy W_o, the unavailable energy W_u, and the dispersed energy W_d is (Fig. 2.9)

$$W_i - W_u = W_o + W_d, \qquad (2.20)$$

where W_d must be zero in an ideal quasi static process. Then the "useful energy" is perfectly conserved. Usually W_d yields heat with lower temperature than the input heat. Even if the input is not heat, W_d is dispersed as heat finally. Therefore, the efficiency of the quasi static process must be 100 %, assuming the unavailable energy remains useful. However, the unavailable energy cannot be the aimed energy, so that we introduce a parameter called the "**coupling constant**" which expresses the strength of the coupling between the input and output obtained energies.

Let $F(x,y)$ be Helmholtz's free energy of the conversion medium, x and y represent the state-variables (intensive or extensive variables which represent the state of the conversion system), then we can expand $F(x,y)$ around the equilibrium state of $x=0$ and $y=0$:

$$F(x,y) = F(0,0) + \frac{1}{2}(a_{11}x^2 + 2a_{12}xy + a_{22}y^2) + \text{- - - -}. \qquad (2.21)$$

New variables defined by

$$X = \sqrt{a_{11}}x,\ Y = \sqrt{a_{22}}y \qquad (2.22)$$

are introduced and substituted into Eq. (2.21) to produce

$$F(X,Y) = F(0,0) + \frac{1}{2}(X^2 + 2kXY + Y^2) + \text{- - -} \qquad (2.23)$$

where we introduced a parameter defined by

$$k = a_{12}/\sqrt{a_{11}a_{12}}. \qquad (2.24)$$

This parameter is nothing but the "**coupling constant**" for which the relationship:

$$k \geq 1 \qquad (2.25)$$

is valid. The equal sign indicates perfect coupling. In the perfect coupling state, there is no unavailable energy.

We shall consider more about the coupling mechanism. Two kinds of generalized forces (intensive variables) f_x and f_y are introduced, corresponding to the generalized displacements (extensive variables) x and y, respectively, where x and y indicate the states before and after the conversion, respectively. Then we can get

$$f_x = \partial F/\partial x, f_y = \partial F/\partial y. \qquad (2.26)$$

When these equations are applied to Eq. (2.21), we have

$$\left.\begin{array}{l} f_x = a_{11}x + a_{12}y \\ \\ f_y = a_{12}x + a_{22}y \end{array}\right\} \qquad (2.27)$$

where the equality $a_{12} = a_{21}$ is called Onsager's reciprocal relationship.

If the x-kind of energy is converted to the y-kind of energy *via* the medium, then the input energy given to the medium is

$$W_x = xf_x = a_{11}x^2 + a_{12}xy \qquad (2.28)$$

and the output energy from the medium is

$$-W_y = -yf_y = -(a_{12}xy + a_{22}y^2). \qquad (2.29)$$

Now, we take the ratio of them:

$$\lambda = -W_y/W_x = -(a_{12}xy + a_{22}y^2)/(a_{11}x^2 + a_{12}xy) \qquad (2.30)$$

The maximum value of the ratio λ with respect to x and y is readily calculated using $\partial\lambda/\partial x = 0$ and $\partial\lambda/\partial y = 0$. When the resulting values of x and y are substituted into Eq.(2,30), we have

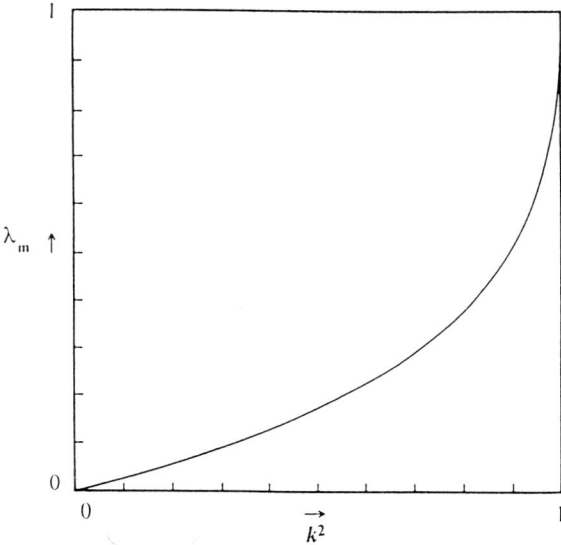

Fig. 2.10. Energy transfer rate and coupling constant.

$$\lambda_m = (1/k - \sqrt{1/k^2 - 1})^2, \qquad (2.31)$$

The maximum value of the ratio λ_m are called the **"energy transfer rate"** which rapidly increases with the coupling constant k increases as shown in Fig.2.10.

Conditions for the possibility of quasi static energy conversion are obviously inferable from Eqs.(2.21) and (2.23), that is to say, if $a_{12} = 0$, then no energy conversion is possible. We can write the condition as

$$\partial^2 F(x,y)/\partial x \partial y \neq 0, \qquad (2.32)$$

which is equivalent to $a_{12} \neq 0$.

If Eq.(2.32) is applied to the three examples described in the previous section, then we have from Eq.(2.8), in the case of Carnot's cycle,

$$\partial^2 W / \partial T_2 \partial V_2 = R/V_2$$
$$\neq 0. \qquad (2.33)$$

In the case of the electric tray, the stored energy in the dielectric medium is given by the formula:

$$W(x,y) = xq^2/(2\varepsilon S), \qquad (2.34)$$

then

$$\partial^2 W / \partial x \partial q = q/(\varepsilon S)$$
$$\neq 0. \qquad (2.35)$$

In the case of the thermo dielectric conversion, we have, from Eq.(2.16),

$$\partial^2 W / \partial V \partial T_2 = Cq/(\alpha S^2)$$
$$\neq 0. \qquad (2.36)$$

A few examples of applied machines of the quasi static process type in practical use are found in internal combustion engines such as the Otto Cycle, the Diesel cycle, the Atkinson cycle, and the Lenoir cycle.

External combustion engines such as the Brayton cycle and Stiring engine are also applications of the quasi-static process.

The conversion between electrostatic energy and the mechanical engine also belongs to this type. The output of the electrical energy from the electrostatic converter is not large (electric current is small) and is commonly utilized in the fields of electronics.

2-3. Thermoelectric Type Conversions

Quasi-static conversion is composed of some processes that are discontinuously undertaken by different types of operations. In order to make a continuous cycle, the working substance should be a fluid flowing in stationary behavior. This offers great advantages in practical applications such as pipe line transportation instead of batch systems.

If the working substance is composed of electrons or ions, the motive force is the electric force or the thermal gradient and the flow velocity is so fast that an ideal conversion such as Thermoelectric Type Conversion is possible.

(a) Thermoelectric conversion

A typical example of this conversion is the thermocouple that converts heat to electricity and *vice versa*. The thermocouple is a direct energy converter and shown in Fig. 2.11. When two kinds of metals, for example, Sn and Bi are connected with each other and one junction is heated, then an electric voltage appears. This phenomenon was discovered by T.J. Seebeck (1770 - 1831) and is named after him.

The resultant voltage on the voltmeter in Fig. 2.11 is expressed by

$$V = S_{12}(T_2 - T_1), \qquad (2.37)$$

where S_{12} is called the thermopower (thermoelectric power) or Seebeck's coefficient between the materials 1 and 2.

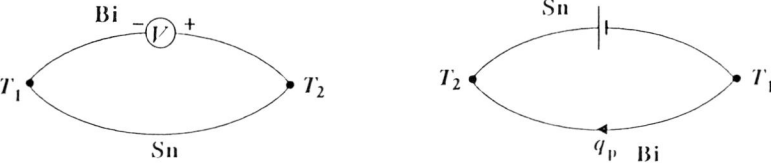

Fig. 2.11. A thermocouple. **Fig. 2.12.** Peltier's effect.

Twenty-three years after Seebeck first discovered the thermoelectric effect, J. Peltier (1785 -1845) discovered a related effect. His circuit is shown in Fig. 2.12. If electric current I is flowing in a closed circuit composing of Sn and Bi wires, then a temperature difference appears. If the direction of the current is reversed, then the direction of the temperature gradient is reversed also, *i.e.*, we have the vector relationship:

$$q_P = \Pi I. \qquad (2.38)$$

$$\Pi = ST. \qquad (2.39)$$

where q_P is the Peltier heat and the proportional coefficient Π is called the "Peltier coefficient". Equation (2.38) means that the Peltier effect is a reverse phenomenon of the Seebeck effect. A rigorous theory has been established by H. Lenz (1804 - 1865) and W. Thomson (1824 - 1907), to describe these and other associated phenomena such as thermo-magnetic effects and the **Thomson's effect** (if an electric current is flowing in a conductor with a temperature gradient, then heat is absorbed or yielded besides the Joulian heat). The most prominent behavior of these thermoelectric effects is that they are reversible processes and different from the Joulian effect. We shall study thermoelectric conversion as a continuous quasi-static effect in detail.

(i) *Basic theory*. We introduce the electric current density I ([A/m^2]) and the entropy flow density S ([J/m^2]), and assume that the thermocouple is composed of a metal with an electric conductivity σ, heat conductivity κ, thermoelectric power S, and thermal gradient $\partial T/\partial x$, coupled with an ideal wire with $\sigma = 0$, $\kappa = 0$, and $S = 0$. Such a thermocouple is called an absolute system of one metal. We have for such a system:

$$\left.\begin{array}{l} I = \sigma E + \sigma S\ (\partial T/\partial x) \\[4pt] S = q/T \\[4pt] = \sigma S E + (\kappa/T + \sigma S^2)(\partial T/\partial x) \end{array}\right\} \quad (2.40)$$

where E is the applied electric field. The thermal gradient and the applied electric field are the motive forces to drive the entropy flow and the electric current. The coefficient of $\partial T/\partial x$ in the right hand side of the first equation is equal to that of E in the right hand side of the second equation. This is due to Onsager's reciprocal relationship. We must notice that the second effect due to Peltier heating, *i.e.*, the entropy flow by the Peltier heat:

$$q_P/T = \Pi I$$
$$= S^2 \partial T/\partial x. \qquad (2.41)$$

This contribution is added to the usual entropy flow by the heat conduction. Comparing Eq.(2.24) with Eq.(2.40), we have for the coupling constant

$$k^2 = ZT/(1 + ZT) \qquad (2.42)$$

with
$$Z = \sigma S^2/k, \qquad (2.43)$$

where Z is called the figure of merit. When $Z \to \infty$, we have $k^2 = 1$, *i.e.*, perfect coupling is possible. The meaning of Z is important in the practical application of thermoelectric power generation described in a later section.

Now, the maximum transfer rate λ_m expressed by Eq. (2.31) is, in the present case,

$$\lambda_m = (1/k - \sqrt{1/k^2 - 1})^2$$

$$= k^2/4 \, (1 - k^2/2 + \cdots), \qquad (2.44)$$

into which Eq.(2.42) is substituted, then we have for the maximum transfer rate:

$$\lambda_m = ZT/[4(1 + ZT)]. \qquad (2.45)$$

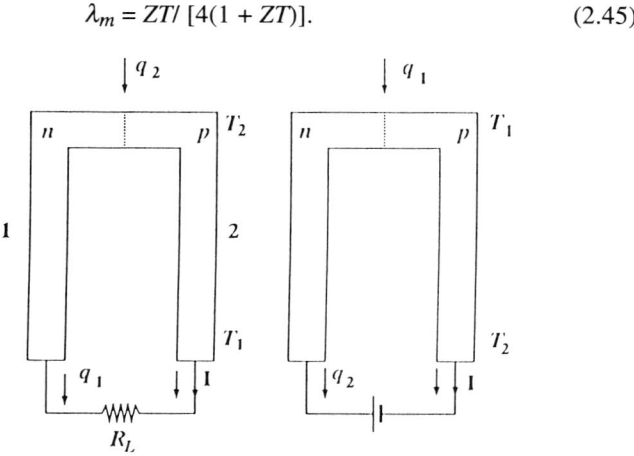

Fig. 2.13. Thermoelectric generator(left) and Peltier refrigerator.

(ii) *Practical application theory.* We shall consider a thermoelectric generator that is composed of *n*- and *p*-type semiconductors (Fig. 2.13). Some of the alloy semiconductors have a thermoelectric power as large as 100 times that of metals. Thermoelectric figure of merit *vs* temperature curves are plotted in Fig. 2.14. For the simplicity, the *n*- and *p*-type semiconductors are denoted by the subscripts 1 and 2, respectively. When the input heat q_2 at the semiconductor junction is separated into the Peltier heat q_P and the conduction heat q_c, we get

$$q_2 + \frac{1}{2} q_j = q_P + q_c, \qquad (2.46)$$

where q_j is the generated Joulian heat, half of which is transported down to the other, free ends of the elements.

The resistance is denoted by R_1 and R_2, then we have

$$q_j = (R_1 + R_2)I^2. \qquad (2.47)$$

The Peltier- and the conduction- heat are expressed by

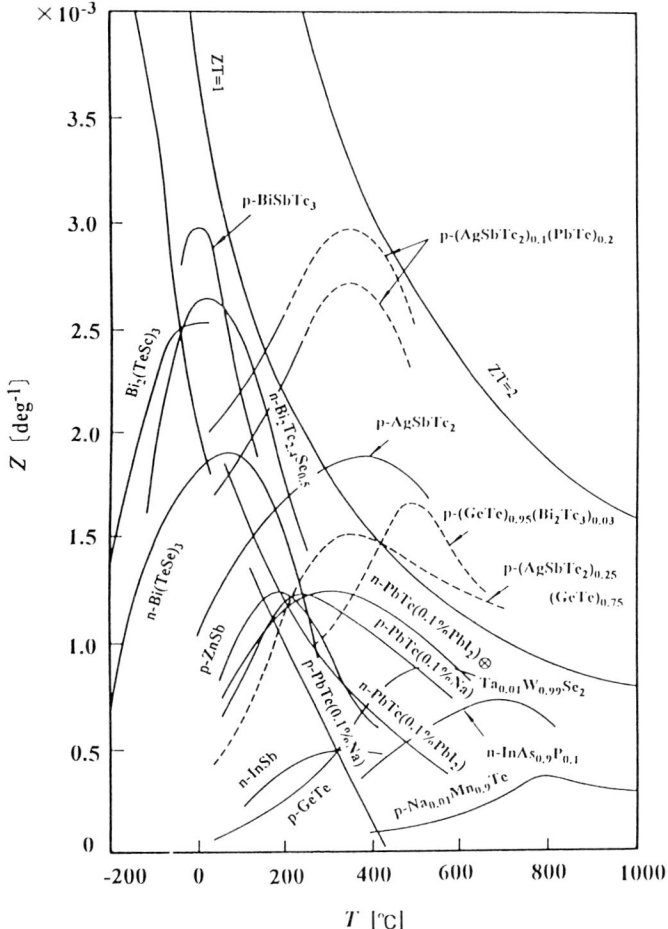

Fig. 2.14. Figure of merit *vs* T °C. T. Okada : *Energy Tomorrow* (an old Japanese magazine)1 (1963) 10.

$$q_P = S_{12}IT_2 \tag{2.48}$$

and

$$q_c = (k_1 + k_2)\Delta T,$$

$$\Delta T = T_2 - T_1, \tag{2.49}$$

respectively, where $R_{1,2}$ and $k_{1,2}$ are the electrical resistances and the thermal conductances, respectively.

Thus we have, for Eq.(2.46),

$$q_2 = S_{12}IT_2 - \frac{1}{2}(R_1 + R_2)I^2 + (k_1 + k_2)\Delta T. \tag{2.50}$$

The work done at the load resistance R_L is maximized under the condition:

$$R_L = R_1 + R_2, \tag{2.51}$$

$$R_1 = R_2, \tag{2.52}$$

and the electric current under the condition is

$$I = S_{12}\Delta T/(2R_L). \tag{2.53}$$

Therefore the work W per unit time is

$$\frac{dW}{dt} = R_L I^2 = (S_{12}\Delta T/2)^2/R_L, \tag{2.54}$$

and the efficiency is given by

$$\eta = W/q_2$$

$$= \Delta T/[\,2T_2 - \Delta T/2 + 4(k_1 + k_2)(R_1 + R_2)/S_{12}^2\,]. \tag{2.55}$$

If the cross section areas, lengths, resistivities, and heat conductivities are denoted by $A_{1,2}, L_{1,2}, \rho_{1,2}$, and $\kappa_{1,2}$ for the elements 1 and 2, then we have

$$(k_1 + k_2)(R_1 + R_2) = (\kappa_1 A_1/L_1 + \kappa_2 A_2/L_2)(\rho_1 L_1/A_1 + \rho_2 L_2/A_2),$$

which can be minimized under the geometrical condition:

$$(A_1/A_2)(L_2/L_1) = \sqrt{\rho_1 \kappa_2/\rho_2 \kappa_1}. \tag{2.56}$$

Therefore we shall design the elements so as to satisfy Eq.(2.56). Now, the efficiency is given by

$$\eta = \Delta T/(\,2T_2 - \Delta T/2 + 4/Z\,) \tag{2.57}$$

where we defined the synthesized figure of merit

$$1/Z = [\,(1/Z_1)^{1/2} + (1/Z_2)^{1/2}\,]^2,$$

$$Z_{1,2} = S_{1,2}{}^2 / \kappa_{1,2} \rho_{1,2}$$

where S_1 and S_2 are the absolute thermoelectric power of the elements 1 and 2. Comparing Eq.(2.45) with Eq.(2.57), we get approximately

$$\eta = (\Delta T/T)(ZT/4)(1 - ZT/2 + \cdots)$$

$$= \eta_c \lambda_m [1 - O(ZT)^2], \qquad (2.58)$$

where η_c is the Carnot's efficiency and the term $O(ZT)^2$ is a term with order of $(ZT)^2$ which is small enough compared to 1. We have a relationship:

$$\eta = \eta_c \lambda_m. \qquad (2.59)$$

(iii) *Entropy production.* Applying Eq.(2.20) we shall show that the efficiency of thermoelectric conversion is, in principle, small because the heat conduction is a typical irreversible phenomenon with entropy dispersion, so that the dispersed heat W_d is relatively large. Next, the input heat is conducted to the other ends of the semiconductor elements mainly by phonon transport.

The electric current is accompanied by an electron flow that is caused by the temperature gradient. However, the density of free electrons in a semiconductor is so small that they cannot carry an appreciable part of the input heat. This means that the unavailable heat W_u due to phonons is relatively large, too.

We shall note that metallic materials cannot be used in practical applications because their thermoelectric powers S are very small compared to semiconductors, although their electrons can carry both heat and electricity at the same time.

The entropy production rate is given by

$$\& = IE + S(\partial T/\partial x) \qquad (2.60)$$

which is calculated to give

$$\& = I^2/\sigma + (\kappa/T)(\partial T/\partial x)^2. \qquad (2.61)$$

Now, we shall examine whether the entropy production rate & can be zero or not in the case of $\sigma \to \infty$ that is realized in the superconducting mixed state (the thermoelectric power S is zero in the superconducting state, but the heat conductivity is not zero). Therefore we get

$$\&_m = (\kappa/T)(\partial T/\partial x)^2. \qquad (2.62)$$

This formula shows that the entropy flow is realized only by heat conduction, however, if we pass an electric current in the reverse direction to the conduction heat flow and yield a Peltier heat flow q_P that cancels the conduction heat, then the entropy production rate is perfectly zero. This phenomenon was first appreciated by de Groot in his well known book[13].

The order of the magnitude of I_p is given by, from Eq.(2.61),

$$I_p = (\kappa/ST)(\partial T/\partial x). \qquad (2.62)$$

Not only perfect coupling but also 100% energy conversion are possible in this case. Such a case is very seldom found in nature. The phenomenon is simply described, that an electric current is flowing in a mixed state of a superconductor, to which also a heat input is given and still no change is observed.

(iv) *Applications*. The merits of thermoelectric conversion are that it has no moving parts and makes no noise. Moreover, these systems are useful to extract the unavailable heat from any heat engine. The electromotor to drive the exhaust fan for a gas stove can be powered by thermoelectric generation utilizing the waste heat from the stove. The cooling fan of a slide projector can also be motivated by a thermoelectric generator. Examples like these are many; this type of generator is applied mainly to electronic and small scale electrical machines.

A famous power generator applied to space vehicles is SNAP (System of Nuclear Auxiliary Power) in U.S.A. The SNAP 10A is a thermoelectric generator with maximum output of 28.5 [V] x 19 [A]. A Ge-Si alloy whose figure of merit is 0.58×10^{-3} [1/K] is applied and the heat source is provided by a small nuclear reactor with a diameter of 22.7 [cm] and length of 39.6 [cm]. 1,440 thermocouple pairs are used, each of which has an open circuit voltage of 0.034[V].

Table 2.3. Thermoelectric refrigerator and Carnot's refrigerating engine (refer Fig. 2.13)

Thermoelectric Refrigerator	Carnot's Refrigerating Engine
Heat conduction $K(T_2 - T_1)$	Gas leak from piston and valve
Joulian heat in elements $1/2\ RI^2$	Friction heat of working fluid, over heat of piston and valve
Thermoelectric power S_{12}	Pumping efficiency that is decided by compression ratio
Figure of merit Z	Refrigeration power per unit volume of piston motion
Maximum temperature difference $\triangle T = 1/2\ ZT_i^2$	Maximum pressure difference

The future of the thermoelectric power generator depends on the development of new thermoelectric materials which have large figures of merit at high temperatures.

(v) *Thermoelectric refrigeration*. The Peltier effect can be applied to a refrigerator. The principle is shown in Fig.2.13 (right hand side). An electric current I carries Peltier heat q_2 (= ΠI) from the p-n junction to the free ends of the semiconductors so that the junction is cooled down and the free ends are warmed up.

We show a comparison of the thermoelectric refrigerator with the conventional Carnot refrigeration engine in Table 2.3.

The merits of this type of refrigeration are the same as those of the thermoelectric generator (no noise, no vibration, and so on). Thermoelectric refrigerators are applied to the cooling of electric and electronic devices and machines, *e.g.*, the cooling box applied to a detection device made of PbS elements, for the infrared rays. The temperature of a developing tray for the photographic film is also kept constant by applying Peltier refrigeration. Applications also occur in the medical and physiological fields, because they supply limited refrigeration. The essentially important quantity of thermoelectric energy conversion is the figure of merit Z, which is plotted against the temperature T in Fig. 2.14.

(vi) *Thermo-ionic power generation*. The electrons in a metal cannot leave the attractive potential of the metal ions. However, if the temperature of metal is so high that the kinetic energy of electrons exceeds the potential energy, which is called the work function ϕ, then the electrons are emitted out. This is the phenomenon called thermoionic emission.

Utilizing this phenomenon, an electrical generator is designed as shown in Fig. 2.15. A and C are the anode and the cathode, respectively, and the space between A and C is evacuated. The cathode C is heated near the temperature $T_c = \phi/k$, where k is Boltzmann's constant, then a thermoionic current flows in the circuit:

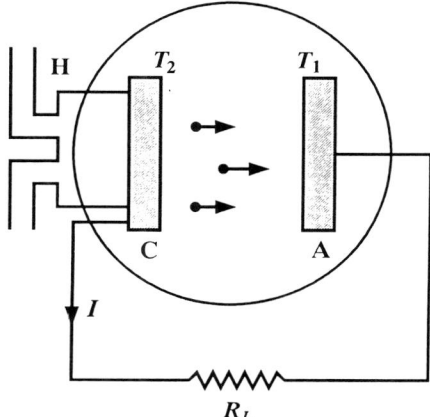

Fig. 2.15. Principle of the thermoionic converter.

$$I_T = DT^2 \exp(-\phi/kT) \qquad (2.64)$$

where D $(= 1.2 \times 10^6$ $[A/(m^2 \cdot K^2)])$ is a constant and Eq.(2.64) is called the Richardson-Dushman formula.

The thermoionic circuit can be regarded as a kind of thermoelectric conversion where one metal is replaced by the vacuum. The vacuum has a heat conductivity so small that the figure of merit becomes very large. On the other hand, another unavailable heat is generated, *i.e.*, the radiation energy expressed by Eq.(1.33). The temperature is so high that heat radiation cannot be neglected. Moreover, the emitted electrons in the narrow space between A and C are stagnated by exclusive interactions between themselves. This produces the resistance of the circuit. The resistivity due to the electron plasma[45] is given by

$$\rho_p = F \ln(\lambda/T^{3/2}) \qquad (2.65)$$

$$\lambda = (3/2e^3)(k^3T^3/\pi n)^{1/2}, \qquad (2.66)$$

where F $(= 65.3$ $[\Omega M])$ is the constant and n is the electron density.

The heat conductivity of the electron plasma is estimated by the Wiedemann-Franz law, which indicates that the ratio of the heat conductivity to the electrical conductivity is constant for metals, and that this constant is proportional to the absolute temperature. We approximate the heat conductivity of plasma to that of a

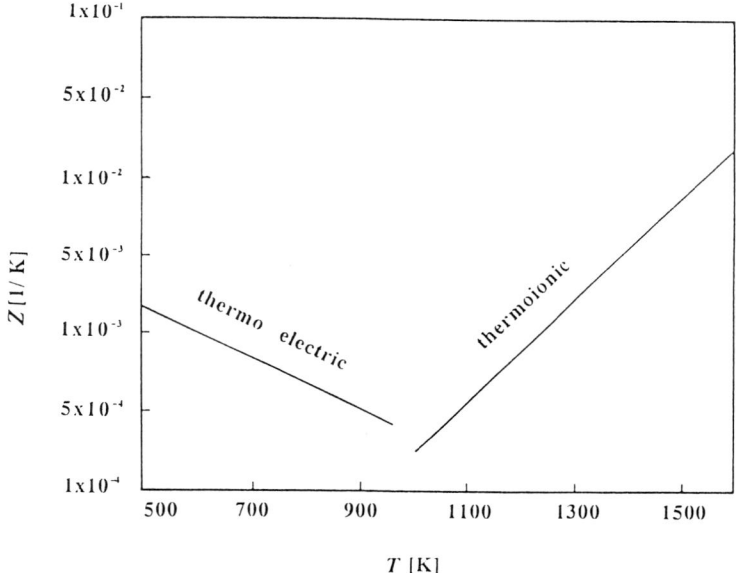

Fig. 2.16. Temperature dependence of figure of merits in thermoionic and thermoelectric conversions.

metal, and get

$$\kappa_p = 2.5 \times 10^{-10} \, T/\rho_p. \tag{2.67}$$

According to A.H. Wilson[49], the thermoelectric power of free electron systems is given by

$$S_p = (k/e)[\, 4 - \ln(nh^3/(2(2\pi mkT)^{3/2}))] \tag{2.68}$$

which can be approximately applied to the electron plasma. The e and m in Eq.(2.68) are the electronic charge and mass, respectively.

Substituting Eqs.(2.65), (2.67), and (2.68) into Eq.(2.43), the figure of merit can be estimated to be of the order of 3×10^{-3} [1/K] and $ZT = 3$ at 1,000 [K]. This is a very high value compared to that of thermoelectric conversion and gives the conversion efficiency of the order of 3 %, keeping the temperature difference between A(2,500 [K]) and C(3,000 [K]) constant (= 500 [K]).

We show the figure of merit Z vs temperature T for this thermoionic system in Fig.2.16.

2-4. Dynamic Conversion

A prominent characteristic of conversion machines between electrical and mechanical energies, such as power generators and electromotors, is that the velocity of the moving parts in the conversion system and the coupling strength between the subsystems will decide the efficiency.

The coupling strength is decided by the material, the geometrical arrangement, and other contrived designs that are different, case by case. However, it is true, in any case, that the velocity of the motion of the moving parts plays an essential role. Such conversions are called "dynamic (energy) conversion".

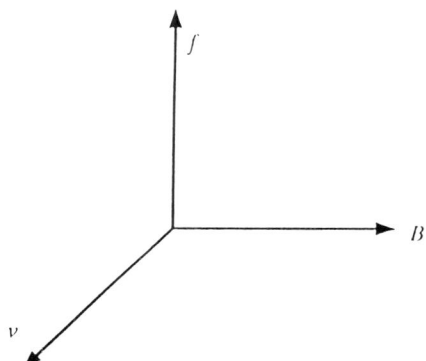

Fig. 2.17. Lorentz's force.

The fundamental interaction between motion and electromagnetic phenomena is due firstly to the Lorentz force, which is shown in Fig. 2.17. A charge q moving with the velocity v along the x-axis in an uniform magnetic field with the flux density B ($[Vs/m^2] = [Wb/m^2]$) along the y-axis, will experience Lorentz's force :

$$f = q[\ v \times B\ ] \qquad (2.69)$$

An electric current density I is written as $I = qv/l$, where l is the length of the leading wire in the system. Therefore the charge q is replaced by the current I:

$$f = l[\ I \times B\ ] \qquad (2.70)$$

If the wire is not linear, then a wire element dl is used, and Eq.(2.70) is replaced by $df = I[\ dl \times B\ ]$.

Another fundamental interaction between mechano-electrical conversion systems is electromagnetic induction. Figure 2.18 shows that an induced current occurs in closed circuit when a magnet is moving (towards or away) from it. The induced voltage is

$$V_i = d\Phi/dt, \qquad (2.71)$$

where Φ ($= SB$, S is the circuit area) is the magnetic flux.

Figure 2.19 shows that an induced current occurs in a closed circuit (b) when the current in the circuit (a) changes. The induced voltage in (b) is also expressed by Eq.(2.71).

Both Eqs.(2.70) and (2.71) mean that rapid changes of q, B, and Φ are essential in the energy conversion. Such systems are typical examples of the present conversion.

Instead of Helmholtz's free energy in Eq.(2.21), the fundamental equation for the energy of this system can be expressed by

$$W = \frac{1}{2}(b_{11}x^2 + 2b_{12}xy + b_{22}y^2\ ) +, \qquad (2.72)$$

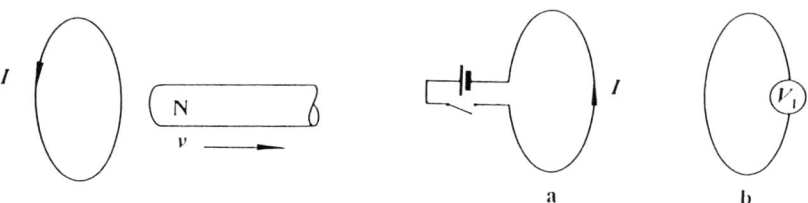

Fig. 2.18. Electric current is induced in the circuit.

Fig. 2.19. Changing electric current induces a current in near circuit.

The term that contributes to the energy conversion is the second term of the right hand side and the coupling constant is defined by

$$k = b_{12}/\sqrt{b_{11}b_{22}}, \qquad (2.73)$$

which is quite similar to Eq.(2.24).

(1) Electric power generator and electromotor
(a) Electric power generator. An electric power generator is a traditional, highly efficient energy converter. The efficiency is very high if the system is designed so as to maximize the coupling constant. The principle of electric generation is as follows. Two coils A and B are placed in series, between which a magnet NS is placed as shown in Fig.2.20. If the magnet is motivated to rotate in the clockwise direction, then an electric current I flows in the direction indicated by the arrow, for the magnet position shown in the figure. This electric current yields a magnetic field that tends to rotate the magnet in the reverse direction. By supplying external energy to the magnet to keep it rotating, the electric current is also maintained. This current is an alternating current, because the direction and the magnitude vary with $\cos q$, where q is the angle between the magnet length and the coil axis.

In order to make the coupling constant large, the turn number density of the coil and the coil shapes are specially designed and a rotor coil is used instead of a magnet.

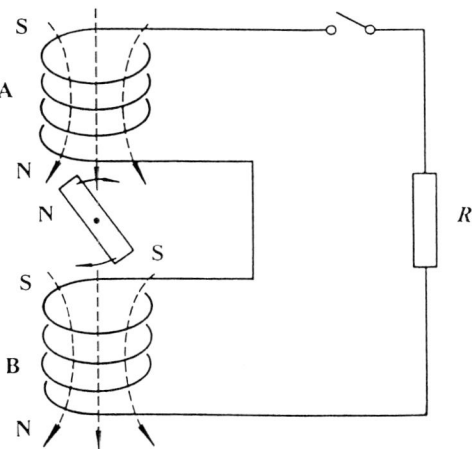

Fig. 2.20. Power generator.

(b) Electromotor. We shall explain the principle of the electromotor using Fig.2.21. When coil is placed among the magnet poles and an electric current is passed, then a torque is generated by the coil, which is then rotated about its axis The torquei s given by

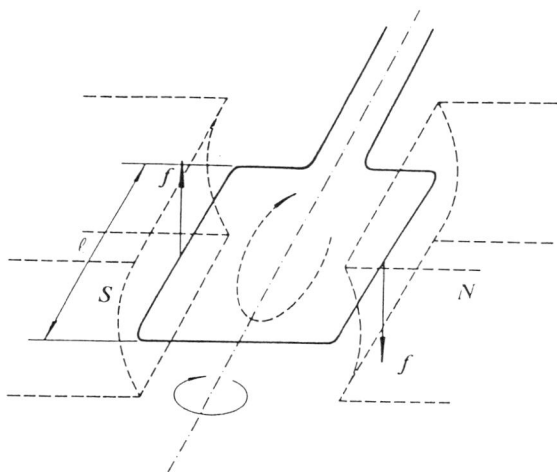

Fig. 2.21. Motor.

$$T = fa \tag{2.74}$$

where f is given by Eq.(2.70) and a is the length of the coil's moment arm and is given by

$$a = d\cos\theta, \tag{2.75}$$

where d is the distance between the two current path of the inserted coil and θ is the angle between the coil area and the plane connecting the center lines of the poles (see Fig. 2.21).

Unavailable energies come from Joulian and frictional heat but these are very small compared to the heat escaped from the heat engines.

(2) Magneto hydro dynamic power generation

M. Faraday invented a generator different from the generator described in the previous section. The structure and the mechanism of his generator is as follows. Imagine a channel where both sides are metal and the bottom is an insulator. A strong magnetic field is applied to the channel in the upward direction. If a fluid conductor such as mercury flows in the channel with high speed, then a voltage will be observed between the side metal plates.

This type of power generator has been recently re-examined. One of the reasons is that the temperature of the heat evolved by the combustion of fossil fuel is as high as 2,000 [K]. However this temperature is too high to be applied to a turbine generator, because the turbine blade cannot withstand a strong and destructive centrifugal force at such high temperatures.

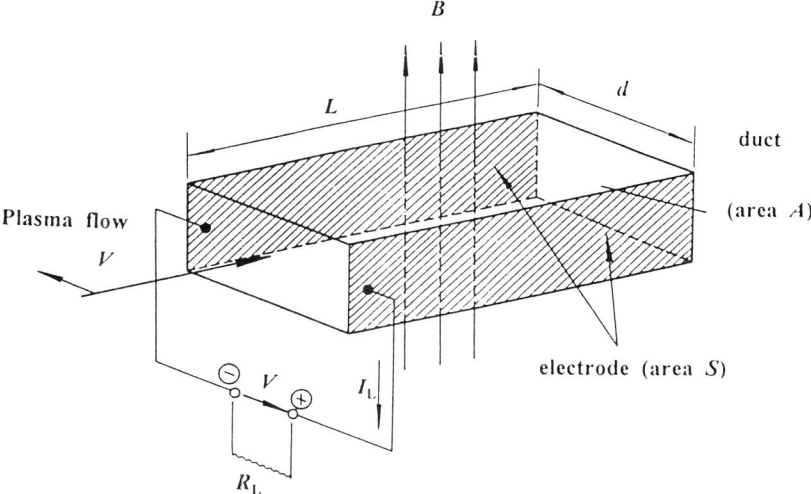

Fig. 2.22. Principle of magnetoplasma-dynamic power generation.

The critical temperature of modern alloys is up to 1,500 [K]. Usually, gas or steam at temperatures from 700 [K] to 1,300 [K] are utilized to power gas- or steam-turbines.

In order to increase the conversion efficiency, we plan to design a generator working at higher temperature so as to achieve a higher Carnot efficiency, Fossil fuels are burned to get a high temperature gas, into which some alkali metals are seeded to provide electrical conductivity, the gas then achieving the plasma state and having conductivity. The plasma gas is pressurized to flow into a duct that is placed in a strong magnetic field B. If both side walls are made of metal, then we can get an open circuit voltage V_0. This principle is shown in Fig.2.22.

We have

$$\left. \begin{array}{l} V_o = vBd \\ R_i = d/(\sigma S) \end{array} \right\} \quad (2.76)$$

where d, S, and σ are the distance between the side walls, the area of the wall, and the conductivity of the plasma gas, respectively.

The loading factor is defined by the ratio of the external load voltage to the generated voltage and is given by

$$\left. \begin{array}{l} \xi = e/(vB) = R_L/(R_i + R_L) \\ E = V_o/d - I_L/\sigma \end{array} \right\} \quad (2.77)$$

where E is the apparent electric field. The loading factor is the ratio of the available electric power at R_L to the electric power converted from the plasma flow.

The output electric power density is defined by the ratio of the available electric power to the duct volume and is given by, from Eq.(2.77)

$$p = \sigma v^2 B^2 \xi (1 - \xi). \qquad (2.78)$$

The power density can be maximized when $\xi = \frac{1}{2}$, i.e., $R_L = R_i$. The maximum power density is

$$p_m = \sigma v^2 B^2 / 4. \qquad (2.79)$$

The only loss (unavailable heat) due to the Joulian heat in the duct is given by

$$p_l = I_L^2/s = \sigma v^2 B^2 (1 - \xi)^2. \qquad (2.80)$$

The conversion efficiency for the maximum power density is given by

$$\eta = p_m/(p + p_m), \qquad (2.81)$$

which achieves a value of 0.5 when p becomes equal to p_m. The maximum value of η is denoted by η_m; it is equal to the maximum value of ξ;

$$\eta_m = \xi_m . \qquad (2.82)$$

The maximum efficiency is 50 % for magnetoplasma conversion efficiency. Another name for **PHD** (Plasma HydroDynamic power generation) is **MHD** (Magneto HydroDynamic power generation).

To avoid losses certain design features are built into the MHD. One of them is the design of the duct form. As the plasma velocity is reduced by the resistance, the duct has a large mouth that becomes narrower and narrower as the distance from the mouth increases in order to keep the plasma speed constant. Another example of an effective design feature is the segmentation of the electrodes. If the electric power is collected at one point on each electrode then the plasma flow lines are distorted, subject to surplus resistance. Moreover, the velocity and the resistivity of the plasma flow in the duct are different depending upon the distance from the mouth. These are the main reasons for the segmentation of electrode, but there is another important reason that is described below (Fig. 2.23).

The **Hall effect** in MHD generation must be taken into account. The Hall effect is explained as follows. An induced electromotive force is exerted upon a

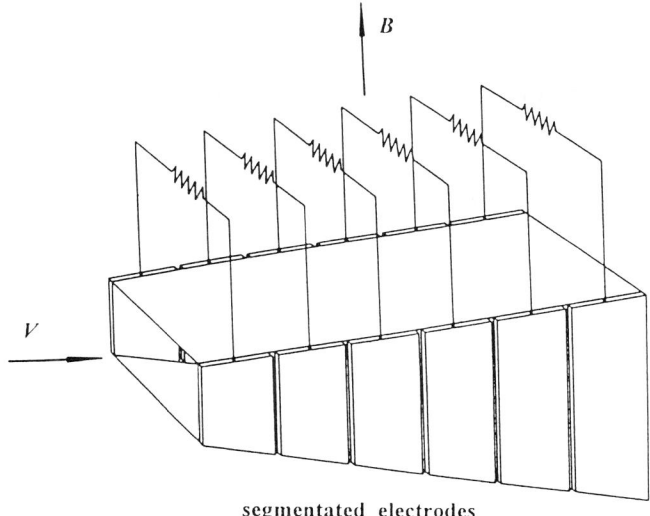

Fig. 2.23. The segmented design of the MHD generator.

charge, in the presence of both electric and magnetic fields, in a direction perpendicular to both fields. Applying this effect to the MHD generator, we have an electric voltage, and accordingly a current, along the plasma motion. This Hall effect is the source of power loss. The loss is estimated by the factor:

$$1/(1 + w_c^2 \tau_c^2) \tag{2.83}$$

where w_c ($= eB/m$) is called the cyclotron frequency (electrons in a magnetic field make a spiral motion; the frequency of revolution around the magnetic flux axis in one second is w_c), and t_c is the time between one plasma particle collision and the next. The scattering particles can be either ions or other electrons.

It is necessary to make $(w_c \tau_c)^2$ very small, but if this is impossible, we must utilize the Hall power, also. The segmentation of the electrodes is useful for gathering this power.

In the 1960s some American companies such as AVCO-Everett, General Electric, Westinghouse, Martin Marietta and some British companies such as CERL, and International Research & Development, and the Japanese governmental Electro Technical Laboratories tried to develop MHD generators with an output of 100×10^3 [kW]. However, innovative alloys with very strong tensile strength at high temperatures have been developed to meet our needs in connection with development of space vehicles. These alloys are now being applied to gas turbines at higher temperatures and we have actually had some applications already. Most gas turbines today have excellent efficiency and no effort has been made in developing MHD generators since the 1980s.

If the higher heat input is effectively utilized by MHD or high temperature gas turbine generators, the wasted unavailable heat can be used again by a steam turbine, so that totally efficient utilization is possible. Such a system is called "topping" and the converter applied at the higher temperature side is called a "topper". This multistage generation will be essential in the future and will be discussed in a later chapter.

(3) Resonant conversion

Let us consider a mechanical vibration system (that is very similar to the oscillating electric current in a damping circuit with an external power source) whose eigen angular frequency is w_0. If an external force with an angular frequency w is applied, then the eigen vibration is damped and only the forced vibration occurs. The equation of motion is

$$\ddot{x} + 2\gamma \dot{x} + w_0^2 x = (F/m)\cos wt, \qquad (2.84)$$

where m is the mass of the vibrator and x represents the displacement from the equilibrium position. The solution of Eq.(2.84) is the real part of the equation:

$$\ddot{z} + 2\gamma \dot{z} + w_0^2 z = (F/m) \exp(iwt) \qquad (2.85)$$

where the new variable z is defined by

$$z = x + iy.$$

Assuming the relationship:

$$z = z_0 \exp(iwt), \qquad (2.86)$$

which is substituted into Eq.(2.85), we have

$$z_0 = (F/m)[(w_0^2 - w^2) + 2i\gamma w] \qquad (2.87)$$

Now the solution of Eq.(2.85) is given by

$$z = (F/m)/\{(w_0^2 - w^2) + (2\gamma w)^2 \exp[i(wt - \alpha)]\} \qquad (2.88)$$

with $\qquad \tan \alpha = (2\gamma w)/(w_0^2 - w^2).$

and the phase is delayed by the angle α. The delayed phase angle is $\alpha < \pi/2$ when $w_0 > w$, and $\pi/2 < \alpha < \pi$, when $w > w_0$. We must notice that $|z|$ becomes maximum

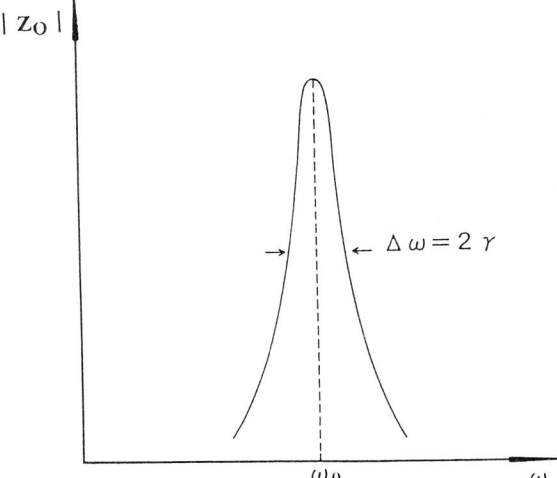

Fig. 2.24. Resonance and half value width.

for $w = w_0$. The maximum value is
$$A_m = F/(2m\gamma w). \tag{2.89}$$

The half value width of the resonant peak is defined by the difference between the angular frequencies at the half value of $|z_0|$ and is given by (Fig.2.24),

$$\Delta w = 2\gamma. \tag{2.90}$$

When the value of γ is small, the peak is high and narrow, because the damping is weak.

This phenomenon is called resonant vibration (or oscillation).

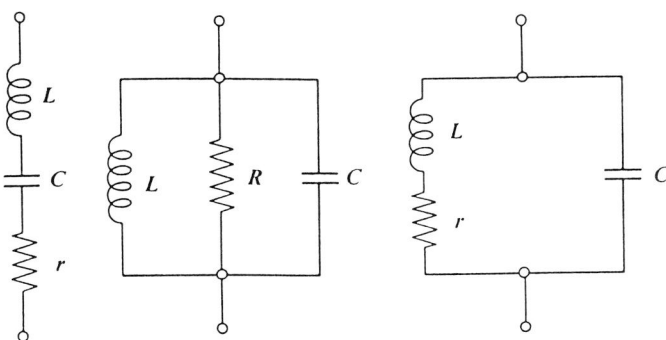

Fig. 2.25. Resonance electric circuits. Series, parallel, and hybrid circuits.

To express the strength of the resonance, we introduce another parameter called the "Q-value" (quality value) that is defined by

$$Q = w_0/\Delta w$$
$$= w_0/(2\gamma). \qquad (2.91)$$

Resonance is an important phenomenon not only in electrical and mechanical engineering but also in physics. We show three typical resonant circuits in Fig.2.25; they are series, parallel, and hybrid systems from the left hand side to the right hand side, respectively.

Among the many applications, some examples are the resonance box of a sound wave, TV, radio, the piezo vibrator, cyclotron resonance, and the cavity resonator of a microwave. We also have many resonant phenomena in physics, such as electron and nucleon scattering by atoms and nuclei, respectively, etc.

From the viewpoint of energy conversion, resonance conversion is very interesting. The unavailable energy (electrical or mechanical) is effectively concentrated in the resonant system by properly designing the value of the eigen frequency and circuitry. If the Q-value is high, then very high conversion efficiency is gained.

Besides the small scale converters of electric and mechanical energy, no large scale application is yet envisioned. However, we may conjecture some practical resonant energy converters to extract the oscillating or vibrating energy from sources such as freeway bridges and the railway base of the bullet train. The electric power generated in such ways can be applied to light the street lamps along the roads and railway.

2-5. Photon Energy Conversions

The study of photon energy conversion has flourished since H.R. Hertz (1857 - 1894) first discovered the photoelectric effect. The photoelectric effect exerted a great influence upon classical physics. The reason is as follows. The kinetic energy of the emitted electrons from a metal depends on the frequency of the irradiated light not on the intensity. This result lead to Planck's discovery of the photon.

When photon energy is irradiated to a material, then the material is warmed if the photon energy is absorbed to transfer its energy to the atoms or electrons in the material. However, the photon is quantum having an energy of $h\nu$ (eg. Section 1-2, (5)). If it excites an electron to a higher energy state so that it is emitted outside the material then the effect is called the "photoelectric effect". This is the external effect and is often called the **external photoelectric effect**. On the other hand, if the electrons are excited but remain inside the material, then it is called the "**internal photoelectric effect**".

In the case of the external photoelectric effect, the electric current I_p is expressed by

$$I_p = AT^2 \phi(x)$$

with
$$x = h(\nu - \nu_0)/(kT)$$

$$\phi = \int_0^x \ln[1 + \exp(x-y)] dy \tag{2.92}$$

where $\nu_0 = W/h$ (W is the work function of the metal), A is decided by the material, and $\phi(x)$ is a known function.

The external photoelectric effect is applied to the photoelectric tube, photomultiplier, and so on.

We shall describe internal photoelectric effects, which are (i) the photovoltaic effect, (ii) photo-electromagnetic effect, (iii) photodiffusion effect (Dember's effect), (iv) photoconductive effect, and (v) photochemical effects. Our study is mainly devoted to the photovoltaic and photochemical effects, but the others are briefly explained in the following.

(1) External photoelectric effects

(a) Photo-electromagnetic effect (PEM effect). If the photons are irradiated, along the x-direction of a sample where a magnetic field is applied along the y-direction, then a voltage appears along the z-direction. This phenomenon is called the photo-electromagnetic effect and the current due to this PEM effect may be observed. However, if the current is not perpendicular to the magnetic field, the sample is subject to a torque, which is called the "photomechanical effect".

(b) Photodiffusion effect. Photons are absorbed at the surface of the materials to release electrons and positive holes (the positive hole is a positively charged particle created as a hole where an electron is removed by the collision with the photon). The density of these particles is higher at the surface than inside. Thus a voltage is observed across the thickness of the sample because the mobilities of the electron and hole are principally different. This is called the "photodiffusion effect", or the Dember effect.

T.S. Moss[30] has calculated the photo-diffusion voltage and found

$$V_d = \frac{[(b-1)fL]}{[\mu_n(n+p)(1+\alpha)]}$$
$$b = \mu_n/\mu_p \tag{2.93}$$

where n and p are the densities of electrons and positive holes, $\mu_{n,p}$ is the mobilities of the electron and positive hole, and f and α are the generating rate of the electron-positive hole pairs and the annihilation rate of the pairs by recombination at the surface, respectively. L is the diffusion length of electron. This effect is observable but it is too small to have any application.

(c) **Photoconductive effect.** Let the increments of the electron and positive hole densities by irradiating photons on a substance be represented by Δn and Δp, respectively. The resultant increment of the electrical conductivity is

$$\Delta\sigma = \sigma - \sigma_0$$

$$= e(\Delta n \mu_n + \Delta p \mu_p), \qquad (2.94)$$

The photoconductive effect does not convert photon energy to electric energy that can be taken out of the material, but it is effectively available for signal conversion from light to electricity. The photons come onto the surface of a photoconductive crystal, to which a bias voltage is applied. Then the voltage sensitively changes and an electric current is obtained. This phenomenon is utilized by the exposure meters of cameras, EE-cameras, and detectors of infra-red or ultra-violet rays.

(d) **Photochemical effect.** The structure change of Norbor-nadine by irradiation of light (p.48) is a kind of photochemical effect. In addition, many kinds of photochemical reactions are known.

(i) *One photon-hydrogen radical forming system.* If light with a wavelength shorter than 0.33[μm] is irradiated to an aqueous solution of a catalyzer X, the following reaction occurs,

$$H_2O + X \rightarrow X^+ + H + OH^-. \qquad (2.95)$$

When the light wavelength is 0.25 [μm], the reaction is

$$X^+ + \frac{1}{2}H_2O \rightarrow X + H^+ + \frac{1}{4}O_2 \qquad (2.96)$$

both of which lead to

$$H_2O \rightarrow \frac{1}{2}H_2 + OH \qquad (2.97)$$

(ii) *One photon-hydrogen and oxygen forming system.* A typical scheme is

$$Z + H_2O \rightarrow ZO + H_2$$

$$ZO \rightarrow Z + \frac{1}{2}O_2 \qquad (2.98)$$

Both the first and the second systems need light with a wavelength of 0.42 [μm]. and Eq.(2.98) leads to

$$H_2O \rightarrow H_2 + \frac{1}{2}O_2$$

(iii) *One photon-hydroxy radical forming system.* A typical scheme is

$$\left. \begin{array}{l} Y + H_2O \rightarrow Y^- + H^+ + OH \\ \\ Y^- + H_2O \rightarrow Y + OH^- + \frac{1}{2}H_2 \end{array} \right\} \quad (2.99)$$

which proceed with light of a wavelength of 0.367 [μm] and 0.254 [μm], respectively.

Professor L.J. Heid[21] of M.I.T. in the U.S.A. proposed a photochemical water decomposition system as early as in 1948. The reactions are

$$\left. \begin{array}{l} Ce^{4+} + \frac{1}{2}H_2O \rightarrow Ce^{3+} + \frac{1}{4}O_2 + H^+ \\ \\ Ce^{3+} + H_2O \rightarrow Ce^{4+} + \frac{1}{2}H_2 + OH^-. \end{array} \right\} \quad (2.100)$$

Both reactions are combined to split water. These reactions proceed by irradiating the solution with light possessing a wavelength of 0.253 [μm]. However the efficiency is very low.

Besides these preliminary reactions, some promising and highly organized photochemical water splitting reactions such as ruthenium complex, rhodium complex, and copper complex systems are studied. Decomposition of water will be discussed in a later chapter as a fundamental step in the development of hydrogen energy systems.

(2) Photovoltaic power generator (solar cell)
(a) Characteristics of solar cells. When a photon incident on a semiconductor surface excites an electron from the filled energy band (valence-bond energy band) to the conduction band, then a positive hole is generated.

If light is regarded as an electromagnetic wave, the energy is not enough to excite the electron to jump to the conduction band, because the energy gap between the bottom of conduction band and the top of the filled band is of the order of 0.5 - 2 [eV] in semiconductors. The energy of light waves never attains these values.

Let us consider a *p-n* junction of a semiconductor, as shown in Fig.2.26. The electrons are excited to the conduction band by an attractive force exerted by the potential slope in the transition region between the *p*- and *n*-regions.

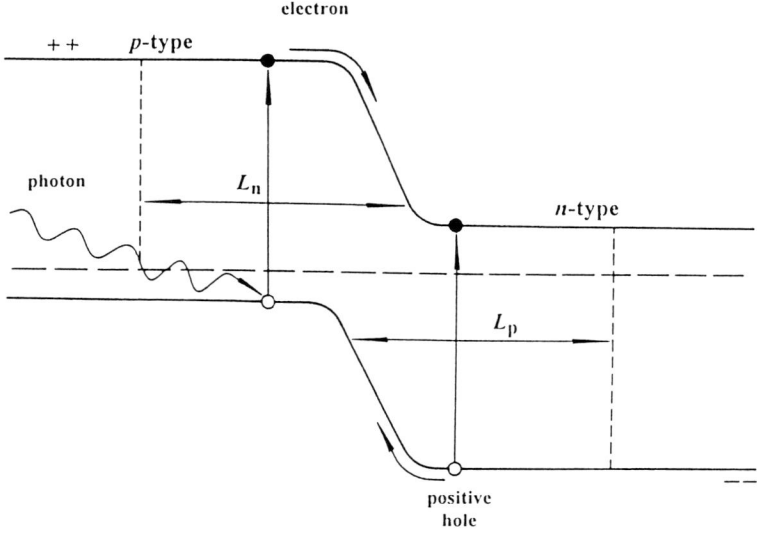

Fig. 2.26. Photovoltaic effect in the p - n junction of semiconductor.

The potential slope is called the **"built-in field"**. The electrons move to the n-semiconductor and the positive holes move to the p-semiconductor, because similar forces acts on the positive holes by the built-in field.

The life times of the electron and the positive hole are denoted by τ_n and τ_p, respectively, and in each characteristic time the carriers travel the distances L_n and L_p, that are called the diffusion lengths.

The positive hole density and the electron density in the n- and p-regions in the thermal equilibrium state are denoted by p_n and n_p, respectively. These are called minority carriers. The saturated electric current across the p-n junction is expressed by

$$I_{so} = e\,(\,L_p p_n/\tau_p + L_n n_p/\tau_n\,). \qquad (2.101)$$

By irradiating the junction with photons, the electron-positive hole pairs are generated at the rate of f per unit time, in unit volume. If the density increase of the minority carriers is $p_n + \Delta p$ and $n_p + \Delta n$, respectively, then we have the following relationship:

$$\Delta p = f\tau_p, \quad \Delta n = f\tau_n. \qquad (2.102)$$

Therefore the electric current across the p-n junction under photon irradiation is

$$I_p = -e(L_p \Delta p/\tau_p + L_n \Delta n/\tau_n)$$
$$= -ef(L_p + L_n). \tag{2.103}$$

This electric current flows from the *n*-region (cathode) to the *p*-region (anode) and generates a voltage in the forward direction. According to this voltage, an electric current I_j is generated in the reverse direction to the above diffusion current which is expressed by

$$I_j = I_{so}[\exp(eV/kT) - 1], \tag{2.104}$$

where V expresses the voltage across the junction.

Thus we can find the net electric current across the *p-n* junction under photon radiation by the sum of $I_j + I_p$,

$$I = I_{so}[\exp(eV/kT) - 1] - ef(L_n + L_p). \tag{2.105}$$

The open circuit voltage is given by $I = 0$:

$$V_0 = (kT/e) \ln[1 + ef(L_n + L_p)/I_{so}]. \tag{2.106}$$

This equation indicates that $V_0 \propto f$, i.e., V_0 is proportional to the radiation intensity, because f is proportional to the radiation intensity. However, if the radiation increases, then f increases also and the left hand side of Eq.(2.106) shows a saturation effect. The maximum voltage of V_0 cannot exceed the band gap energy of the semiconductor. One should also notice that the electrons and positive holes that contribute to the photovoltaic effect must be in the range of $L_n + L_p$ around the junction.

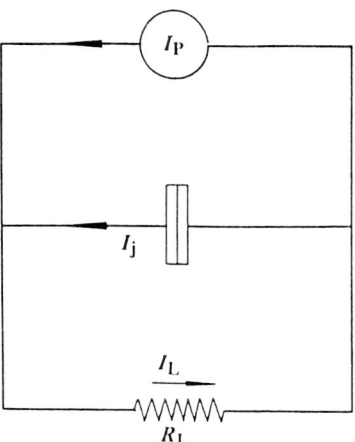

Fig. 2.27. Equivalent circuit.

Next, we shall consider the characteristics of the case where a load resistance R_L is included. The equivalent circuit is shown in Fig.2.27. If the cross-sectional area of the p-n junction is A and voltage V is generated between both ends of the junction under the irradiation of photons, the following three currents exist

(i) *Current through the load*: $I_L = V/R_L$.

(ii) *Forward current*: $I_j = I_{so}A[\exp(eV/kT) - 1]$.

(iii) *Photon induced current*: $I_p = -efA(L_n + L_p)$.

When these are applied to the relationship :

$$I_L = I_p + I_j,$$

we get

$$V_L/R_L = I_L$$

$$= A[I_{so}(\exp(eV/kT) - 1) - ef(L_n + L_p)]. \quad (2.107)$$

When $V = 0$, then the equation $I_L = efA(L_n + L_p) = I_p$ is true, which shows that I_p is the current in the case of $R_L = 0$, i.e., I_p is the short circuit current.

As the above analysis shows, the characteristic of a solar cell depends on the temperature. At high temperatures, the factor eV/kT and $L_{n,p}$ become so small that the characteristic of I_L vs V becomes worse.

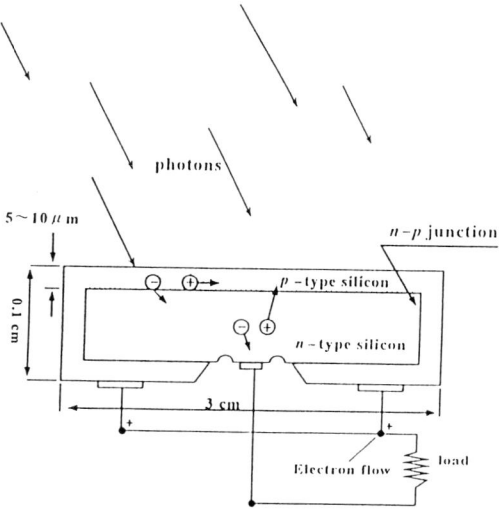

Fig. 2.28. A typical solar cell.

(b) Structure of solar cells. Silicon is one material commonly used as the base material in solar cells. Silicon is doped with phosphorus(P) to produce n-type and with boron(B) to produce p-type semiconductors. Recently, CdS, InP, GaAs, CdTe, AlSb, and other semiconductors were studied for use in practical applications.

Semiconductors are applied in single crystal, polycrystalline, and amorphous states. Single crystals are expensive, while amorphous ones have low efficiency. Usually, the polycrystalline materials are most popular and have efficiencies of up to 10-12 % when applied as solar cells.

The typical structure is shown in Fig.2.28 where a p- on n-type Si solar cell is pictured. The sizes shown here are typical. An anti-reflective film coats the surface of the p-Si. Cooling effects must be taken into account when solar cells are applied to practical generation systems.

As mentioned on p.34, the most advanced solar cell has a highest efficiency of 35.6 % under an illumination 350 times that of the collected solar beam. This solar cell is a tandem type (stacked structure) and composed of Ga-Sb coupled with GaAs semiconductors.

Another **tandem type** solar cell is shown in Fig.2.29[43]. The grid electrodes are attached to the surface which the solar beam irradiates, under which p-GaAs is pasted to an anti-reflective film. To make such a stacking with the use of GaAs and GaSb is very difficult and was first made by the Boeing group (p.34).

This type of solar cell was first manufactured in 1988 at the Sandia National Laboratories in the U.S.A. and it has the efficiency as high as 31 % under illumination of 350 times of the solar beam. The top cell and the bottom cell have 27.2% and 3.8 % efficiencies, respectively.

Lastly, one must be aware that high efficiency does not necessarily mean low cost electric power generation because the price of the cells themselves and the associated equipment must be taken into account. Collection of the solar beams is also a necessary expense. We shall consider cost analysis in a later chapter.

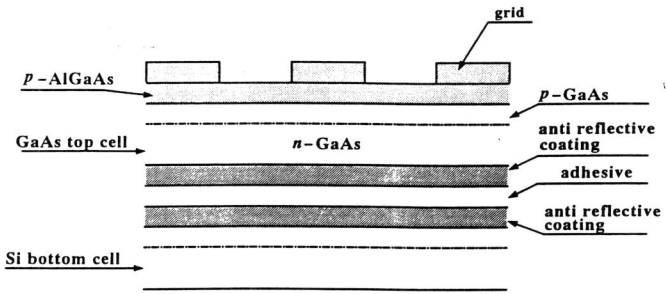

Fig. 2.29. Tandem type solar cell. The materials indicated in the figure were firstly utilized by Sandia National Laboratories in the U. S. A.

Chapter 3
Evaluation of Energy

Iron manufacturing

Chap. 3. Evaluation of Energy

Five types of energies were classified in Chapter 1. Now we shall consider the quality of energy, and which energy has the highest quality. If energy A is relatively easier to convert to energy B but energy B is relatively harder to convert to energy A, then the quality of energy A is defined as being higher than that of B. The ranking of energy quality is also defined in a similar way.

This classification can be easily understood when considering the same type of energy, for example, the energy of a moving body has higher quality when it moves with higher velocity, heat with higher temperature has higher quality, light with higher frequency has higher quality, and so on.

Similar concepts are applied to classify the five types of energies:

(i) Conversion between mechanical energy and electrical energy is achieved with very high efficiency. The turbine generator and the electromotor have nearly the same efficiency, therefore we cannot say which has the higher quality. However, the only electrical energy which exists in natural circumstance is lightning, while many mechanical energies exist. Therefore electromagnetic energy is ranked first, followed by mechanical energy.

(ii) Photon energy can be produced by electric current, but this process requires heat with high temperature. Mechanical energy can hardly be converted to photon energy, but photons exert mechanical pressure upon the comets. However, this condition is not true in terrestrial circumstances and we may say that mechanical energy is still of higher ranking on earth. Moreover, photons can be converted very easily to heat. Photon energy is therefore, ranked third.

(iii) Any kind of energy eventually becomes heat with low temperature and hence heat energy has the lowest ranking. The higher the temperature of heat, the higher the quality of energy. The critical temperature at which the quality of heat exceeds those of solar photons and electrical energy is estimated as follows. If the average photon energy E_s satisfies the condition

$$E_s = kT_c,$$

then heat with a higher temperature than T_c is ranked above the solar photon. E_s is equal to about 1 [eV], so that $T_c = 2,650$ [K].

(iv) Insofar as chemical energy is concerned, it can be converted easily to heat, but is difficult to change to other kinds of energies and is therefore, ranked before heat.

The order of the **energy quality ranking** is thus:

(1) Electromagnetic (2) Mechanical (3) Photon (4) Chemical (5) Heat

Another scientific analysis of energy ranking is the concept of "availability" that was proposed by the late W. Thomson in 1851. His scientific investigation was continued by Z. Rant[33,42] in Germany and developed as "die Exergie" (the exergy). This concept indicates that the available energy can be estimated by taking its circumstances into account. For example the availability of photon energy in space is much higher than that on earth.

Secondary energy was also discussed in Chapter 1. Primary energy is manufactured by many processes to obtain secondary energy, the price of this energy estimated by corresponding procedures. How is the price estimated? We shall show a clear formula for answering this question in the second section of this chapter.

3-1. Exergy

(1) Introduction of exergy

In the introductory paragraph, we learnt about the quality ranking of energy. However, that ranking does not always coincide with the actual value of energy, for example, the temperature of a refrigerated room has more value than the ambient temperature. This value is sometimes called a "negative value". The concept of exergy is the technical term, by which the negative value is scientifically systematized. We may understand the cooling value by the work that is needed to pump heat out from the room.

A similar example is the evacuation value of a gas cylinder. Figure 3.1 shows an evacuated gas cylinder absorbing the air from the outside while the air jet drives an air turbine. A vacuum has less pV energy (p is the pressure and V is the constant volume of the cylinder) if p is lower. Nevertheless, it can do work.

Both examples show that the value of the energy depends on the circumstance, *i.e.*, the parameter of the circumstance is the temperature and pressure in the former and latter cases, respectively.

The value of the room cooling comes from the "produced cost" needed to obtain the refrigeration state, while the value of the vacuum turbine is due to the work produced by the state and is the "available value".

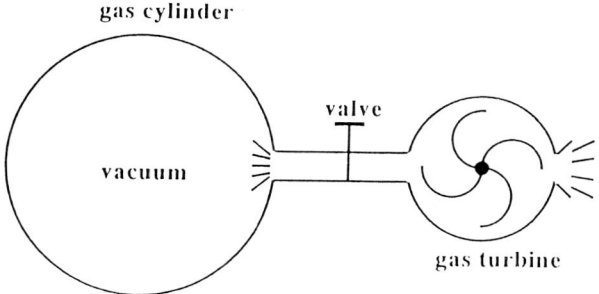

Fig. 3.1. Vacuum turbine.

92   Energy Technology

(a) **Heat exergy.** Taking a heat medium of n mol with the molar specific heat at constant pressure C_p, the heat capacity of this medium is nC_p. The heat of this medium can do the work given by

$$dW = \eta_c \, dQ$$

$$= \eta_c \, CdT, \quad C = nC_p \quad (3.1)$$

where $CdT = dQ$ is the heat possessed by the medium and the Carnot efficiency is $\eta_c = (T - T_0)/T$, where T_0 is the environmental temperature.

The exergy of this heat medium is.

$$E = C \int_{T_0}^{T} dT - CT_0 \int_{T_0}^{T} \frac{dT}{T}$$

$$= C(T - T_0) - C[T_0 \ln(T/T_0)]. \quad (3.2)$$

When the magnitude of $T - T_0$ is small compared with T and T_0, we get from Eq.(3.2),

$$E = C/(2T_0)(T - T_0)^2 \quad (3.3)$$

where we must notice that the temperature difference appears as its square, so that the exergy does not depend on whether T is higher than T_0 or not. The value of heat is proportional to the square of the temperature difference between the heat medium and its circumstance.

Now the heat capacity is defined by introducing the enthalpy H through the relationship;

$$C = dH/dT \quad (3.4)$$

Combining this equation with Eq.(3.2), we get

$$E = H(T, p_0) - H(T_0, p_0) - T_0 \int_{T_0}^{T} \frac{dT}{T} \quad (3.5)$$

where dH/T in the last term of the right hand side of the above equation is the differential entropy dS, i.e.,

$$S = \int dS$$

$$= \int_{T_0}^{T} \frac{dH}{T}. \tag{3.6}$$

Then we have, for Eq.(3.5),

$$E = H(T, p_0) - H(T_0, p_0) - T_0[S(T, p_0) - S(T_0, p_0)] \tag{3.7}$$

A change of a medium volume from V_0 to V can do work against its circumstance (if $V > V_0$)

$$p_0 (V - V_0),$$

which is added to the internal energy U and leads to the enthalpy

$$H = U + pV, \tag{3.8}$$

Substituting Eq.(3.8) into Eq.(3.7), we have

$$E = U(T, p_0) - U(T_0, p_0) - T_0[S(T, p_0) - S(T_0, p_0)]$$

$$+ p_0(V - V_0). \tag{3.9}$$

The above equation is the general expression for exergy. The thermodynamic quantities with subscripts 0 represent environmental conditions.

(b) Pressure exergy. A vacuum's capacity for doing work was mentioned in the introduction to this chapter. We shall now consider it further. If the pressure of the medium is higher than that of its environment, the medium can do the work given by

$$E = \int_{V}^{V_0} (p - p_0)dV, \tag{3.10}$$

because the medium must do the work against the atmospheric pressure p_0, V_0 in Eq.(3.10) expresses the volume of the medium with the pressure p_0.

Applying Eq.(3.10), we have

$$E = nRT_0 \int_{p_0}^{p} (1/p - p_0/p^2)dp$$

$$= nRT_0 [\ln(p/p_0) - (1 - p_0/p)], \tag{3.11}$$

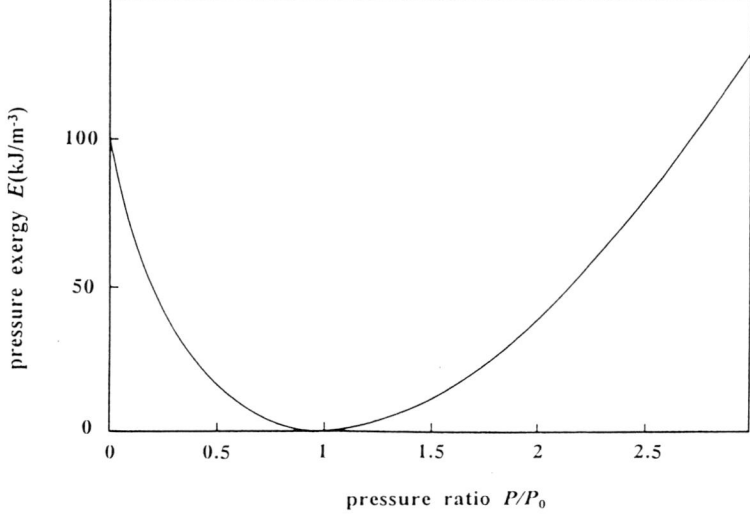

Fig. 3.2. Specific pressure exergy of ideal gas.

where we applied the equation of state for the ideal gas:

$$pV = nRT.$$

Here, we must notice that the exergy E is always positive and is independent of $p > p_0$ or $p < p_0$. Expanding the $\ln(p/p_0)$, we get

$$E = (nRT_0/2) \, [(p - p_0)/p_0]^2. \tag{3.12}$$

The exergy of a gas medium is proportional to the square of the pressure difference under constant temperature, that is, it is the same as heat exergy's dependence on the temperature difference.

The specific exergy is defined by $e = E/V$, i.e., the exergy per unit volume, then we have

$$e = p \ln(p/p_0) - (p - p_0), \tag{3.13}$$

which is plotted in Fig.3.2.

The pressure exergy needed to produce a 1 [m³] vacuum under atmospheric pressure is $e = p_0 = 101.33$ [kJ/m³], when the pressure exergy has a value of $p/p_0 = e \, (= 2.71828)$ at the higher pressure side as shown in the figure.

(c) Latent heat exergy. A phase change between the solid, the liquid, and the gaseous states at constant temperature absorbs or evolves heat, which is called the latent heat.

If the latent heat q_r is taken into a heat medium of temperature T, then the increment of heat exergy is

$$E = q_r(1 - T_0/T) \tag{3.14}$$

The heat exergy must be added to Eq.(3.2) when the temperature changes so widely that the medium is subject to phase changes.

(d) Exergy for the flowing system (pipeline system). Exergy analysis is very useful in the field of chemical engineering. The exergy of a flowing system, *i.e.*, a pipeline system, has a different expression from a closed system in a vessel, *i.e.*, a batch system.

If a medium, which is supposed to be incompressible, is flowing out from a pipeline with a pressure p and a temperature T, then the medium possesses a pressure head of $p - p_0$. This pressure head has the ability of doing the work $(p - p_0)V_0$, therefore we have, for the flowing system,

$$E_f = E_b + (p - p_0)V. \tag{3.15}$$

If Eq.(3.9) is substituted into Eq.(3.15) instead of E_b, the exergy of the batch system, we get

$$E_f = U - U_0 - T_0(S - S_0) + p_0(V - V_0) + (p - p_0)V$$

$$= H - H_0 + T_0(S - S_0), \tag{3.16}$$

where $H = U + pV$ is the enthalpy, *i.e.*, H is the flow of energy divided by the flowing quantity. H_0 is the enthalpy at the environmental state of $p = p_0$ and $T = T_0$. Equation (3.16) is a simpler expression than Eq.(3.9).

(e) Exergy of an open system. An open system is defined as a system into which a medium can flow from the external world. In the case of the pipeline described above, the chemical components of the flow-in medium are assumed to be constant, but, in the case of an open system, the chemical components may change. In such a case, the thermodynamical variables are not only the external variables $V, S...$, but also another external variable $n_1, n_2, n_k...$ that expresses the molar quantities of the component substances of the medium.

The increment of internal energy is then given by

$$dU = TdS - pdV + \sum_{i=1}^{k} \mu_i dn_i \tag{3.17}$$

The first and the second terms of the right hand side of Eq.(3.17) represent the heat increment given from the external system and the work given to the external system, respectively, and the third term is the change in the chemical energy. The chemical energy of the i-th kind of the material is defined by (chemical potential μ_i

x molar number n_i (ref.p.9).

Then we have, for the differential exergy, from Eq.(3.17),

$$dE = (T - T_0)dS - (p - p_0)dV + \sum_{i=1}^{k} (\mu_i - \mu_{i0})dn_i. \quad (3.18)$$

By integrating this equation, we get

$$E = U - U_0 + p_0(V - V_0) - T_0(S - S_0) - \sum_{i=1}^{k} \mu_{i0}(n_i - n_{i0}), \quad (3.19)$$

where we have applied the relationship:

$$U - U_0 = T(S - S_0) - p(V - V_0) + \sum_{i=1}^{k} \mu_{i0}(n_i - n_{i0}), \quad (3.20)$$

where n_{i0} and μ_{i0} represent the molar number and the chemical potential of i-th component material at the environmental state. The molar number n_{i0} is defined at the state of (p_0, T_0, μ_{j0}) where μ_{j0} is realized at the boundary between the system and external environment, for example, a semi-permeable membrane between salt water and pure water has its proper value of μ_{i0}. Copper ferrocyanide precipitated in the fine holes of an unglazed ceramics passes only pure water, not salt water.

In the case of a pipeline system, we have, instead of Eq.(3.16),

$$E = H - H_0 - T_0(S - S_0) - \sum_{i=1}^{k} \mu_{i0}(n_i - n_{i0}). \quad (3.21)$$

(2) Energy conversion and exergy.

Energy conversion has been discussed in Chapter 2. Now we shall consider the relationship between energy conversion and exergy. Let us consider a static (batch) system. The internal energy increment when a heat quantity dq ($= TdS$) comes into the medium is

$$dU = TdS - pdV,$$

where the second term pdV is the work done by the system against the external environment. From this equation, we have

$$T = (\partial U/\partial S)_V, \quad p = -(\partial U/\partial V)_S. \quad (3.22)$$

These relationships are substituted into Eq.(3.9) to obtain

$$E = U(V, S) - U(V_0, S_0) - (\partial U/\partial S)_{V_0,S_0}(S - S_0) + (\partial U/\partial V)_{V_0,S_0}(V - V_0).$$
(3.23)

If the terms containing $(\partial U/\partial V)_{0,0}$, $(\partial U/\partial S)_{0,0}$, and $U(V_0,S_0)$ are subtracted from $U(V,S)$ expanded in the Taylor series around (V_0,S_0), then E in Eq.(3.23) is the rest which is composed of higher than second order terms.

A more general expression is given by analogy to Eq.(3.23),

$$E = U(x_1, x_2, \cdots, x_n) - U(x_{10}, x_{20}, \cdots, x_{n0})$$
$$- \sum_{i=1}^{k} \left(\frac{\partial U}{\partial x_i}\right)_0 (x_i - x_{i0})$$
(3.24)

where $x_1, x_2, \cdots, x_n$ are the extensive variables representing the state of the medium and the suffix 0 indicates the state variables at standard environmental conditions, i.e., $x_1 = x_{10}$, $x_2 = x_{20}, \cdots, x_n = x_{n0}$.

The next term after the third on the right hand side of Eq.(3.24) is the biggest term among the remaining terms in Taylor's expanded U and is written, neglecting the higher order terms, as

$$E = \frac{1}{2} \sum_{i=1}^{k} \sum_{j=1}^{k} \left(\frac{\partial^2 U}{\partial x_i \partial x_j}\right)_0.$$
(3.25)

The exergy E is always a positive quantity so that the matrix whose elements are $\partial^2 U/\partial x_i \partial x_j$ is always positive. Equation (3.25) has a similar form to Eq.(2.21). The discussion on the pages associated with this equation, the coefficient of the second order differentiation of the energy with respect to the extensive variables x and y, proves to be vitally important. Considering such an analogy, exergy conversion is closely related to energy conversion.

The energy conversion efficiency is defined by

$$\eta = \text{output energy / input energy}.$$

The definition given above has, sometimes, an unreasonable result. For examples, the efficiency of a refrigerator becomes negative and the efficiency of heat pump exceeds 100 %.

The exact estimation of energy conversion efficiency should be done using an exergy evaluation by

$$\eta_E = \text{available exergy / input exergy}$$

In order to carry out the calculation of η_E, we need an exergy evaluation for a wider range of physical and chemical quantities. The exergy of substances such as fossil fuels, nuclear fuels, the exhausted gas from a stove, and so on, should be evaluated.

The difference between η and η_E is not small. Some examples follow.

(i) *Electric heater*. The efficiency of an electric heater is considered to be 100 % as all the output becomes heat. However, if the heat exergy is estimated by Eq.(3.3) and if the room temperature increases by 20 [K] from 300 [K] because of the heater, then we have

$$E = C(T - T_0)^2/(2T_0) = (T - T_0)Q/(2T_0)$$

and

$$\eta_E = (T - T_0)/(2T_0) = 1/30.$$

When a fossil fuel is burned to heat up a room, then some substances that have exergy like CO_2 gas are exhausted, so that the exergy efficiency becomes less than that of the electric heater. The exergy of fossil fuel and other chemical substances are discussed in the next section.

(ii) *Heat exchanger*. A heat exchanger transfers the high temperature heat generated by the combustion of fuel to metallic fins, which conduct the heat to heat emitting fins at the lower temperature side.

The heat emitting fins warm up the circulated air in the room. The generated temperature from the fuel combustion is about 1,500 [K] and the exhausted heat from the heat exchanger has a maximum temperature of about 330 [K]. Through the processes, much exergy escapes and the efficiency becomes very small.

(iii) *Combination of fuel cell and heat pump*. If electrical energy is generated by a fuel cell with high efficiency and the obtained electric energy is applied to a heat pump to raise the temperature of a medium, then very high exergy efficiency is realized. Moreover, if the exhausted heat from the fuel cell is also utilized as a heat energy source, then the efficiency is maximized.

(3) Exergy and density change

We shall examine the exergy of a system where the i-th kind of substance comes in or goes out. The molar quantity of the i-th kind of substance and its increment are denoted by n_i and dn_i, respectively.

Now we define the quantity expressing the change rate of exergy with the mol number under T = constant and p = constant.

We define

$$\overline{E_i} = (\partial E/\partial n_i)_{T,p,n_j} \tag{3.26}$$

which is called the partial mol quantity.

The reverse relationship is

$$E = \sum n_i \overline{E_i}$$

and the corresponding thermodynamic quantities are written as
(i) *partial mol internal energy*:

$$\overline{U_i} = (\partial U/\partial n_i)_{T,p,n_j} \quad , \quad \overline{U_{i0}} = (\partial U_0/\partial n_{i0})_{T,p,n_j} \tag{3.27}$$

(ii) *partial mol entropy*:

$$\overline{S_i} = (\partial S/\partial n_i)_{T,p,n_j} \quad , \quad \overline{S_{i0}} = (\partial S_0/\partial n_{i0})_{T,p,n_{j0}} \tag{3.28}$$

(iii) *partial mol volume*:

$$\overline{V_i} = (\partial V/\partial n_i)_{T,p,n_j} \quad , \quad \overline{V_{i0}} = (\partial V_0/\partial n_{i0})_{T,p,n_{j0}} \tag{3.29}$$

If E is written after Eq.(3.19) we have

$$\overline{E_i} = \overline{U_i} - \overline{U_{i0}} - T_0(\overline{S_i} - \overline{S_{i0}}) + p_0(\overline{V_i} - \overline{V_{i0}}) - \mu_{i0}(1 - n_{i0}/n_i)$$

$$= \mu_i(T_0, p_0) - \mu_{i0}(n_i - n_{i0})/n_i, \tag{3.30}$$

where the definition of m_i is applied.

We shall show another expression for the partial mol exergy[24]. The $\overline{E_i}$ defined by Eq.(3.26) can be rewritten as, with p and T constant,

$$\overline{E_i} = (\partial E/\partial n_i)_{T,p}$$

$$= (T - T_0)(\partial S/\partial n_i)_{T,p} + (\mu_i - \mu_{i0})$$

$$= (T - T_0)\overline{S_i} + (\mu_i - \mu_{i0}), \tag{3.31}$$

where $\overline{S_i}$ is the partial mol quantity of entropy and has the relationship

$$\overline{S_i} = -\partial \mu_i/\partial T. \tag{3.32}$$

Now we introduce the concept of "activity" a that is defined by

$$\mu_i - \mu_0 = RT \ln a_i. \tag{3.33}$$

Comparing Eqs.(3.32) and (3.33), $\overline{S_i}$ is written as

$$\overline{S_i} = -R \ln a_i - RT(\partial \ln a_i/\partial T) + S_i, \tag{3.34}$$

where S_i represents the entropy of the i-th kind of substance. When this equation is substituted into Eq.(3.31), we get

$$\overline{E}_i = RT_0 \ln a_i - RT(T - T_0)(\partial \ln a_i/\partial T) + E_i$$

$$= E_i + RT \ln a_i - (1 - T_0/T)h_i, \tag{3.35}$$

where we introduce the heat of dissolution, h_i defined by

$$h_i = RT^2(\partial \ln a_i/\partial T). \tag{3.36}$$

The heat of dissolution is the heat evolved when a solid or a gaseous substance is dissolved into a liquid. The heat of mixing is the heat evolved when a liquid or a gaseous substance is mixed with another liquid or gas. We have one more similar kind of heat that is called the heat of dilution. The heat of dilution is evolved when a solute is mixed with a solvent.

Most textbooks published so far neglect the term h_i for the reason that h_i is too small. However, it can never be neglected in cases where the mixing of sulfuric acid or a freezing mixture is treated.

Equation (3.35) is a convenient formula, because the parameter h_i can be measured by a simple experiment.

(4) Exergy of substances (chemical exergy)

Chemical energy is defined as an affinity energy (cohesive energy) in Chapter 1. **Chemical exergy** is defined likewise and is exactly equal to the chemical energy in so far as it is confined within the substance. However, exergy is released outside the substance whenever a threshold ignition energy is given to it, and heat, entropy, and chemical products are produced simultaneously. Accordingly, chemical exergy escapes, associated with the entropy and the chemical products.

The second definition of chemical energy in Chapter 1. is the density difference energy. This kind of chemical exergy is treated in the same way. We shall introduce some typical examples in this section.

(a) Rich air (air with more oxygen than the atmospheric air).[38]

Equation (3.11) expresses the pressure exergy in a closed vessel with pressure p, placed in an environment with p_0 and T_0. The standard state of the atmospheric environment at the earth's surface is defined by

$$T_0 = 298.15 \text{ [K]}, \quad \text{and} \quad p_0 = 1.01325 \times 10^5 \text{ [Pa]}.$$

Air is composed of many component gas molecules, each of which is indicated by the suffix i.

It is well known that the volume percentages of nitrogen (N_2) and oxygen (O_2) are 78.084 % and 20.948 %, respectively, and the sum of the remaining gases such as Ar(0.934), CO_2(0.033), Ne(0.00182), etc. is below 1 %. Accordingly, we will assume that the components of air are $N_2 = 79$ % and $O_2 = 21$ %. Nitrogen and oxygen components are indicated by subscripts 1 and 2.

The exergy of partial pressure of gases 1 and 2 can be written, using Eq.(3.11), as

$$E_i = n_i RT \,[\, \ln(p_i/p_{i0}) - (1 - [p_{i0}/p_i])\,]$$

$$= n_i RT \,[\, \ln(n_i/n_{i0}) - (1 - [n_{i0}/n_i\,])\,], \qquad (3.37)$$

If new variables are defined by

$$x_i = n_i \Big/ \sum_i n_i \quad \text{and} \quad x_{i0} = n_{i0} \Big/ \sum_i n_{i0} \qquad (3.38)$$

then the relationship;

$$\sum_i n_i = \sum_i n_{i0} \qquad (3.39)$$

is maintained.

Equation(3.37) is rewritten, with use of the new variables, as

$$E_n = n_i RT \,[\, \ln(x_i/x_{i0}) - [1 - (x_{i0}/x_i)]\,]. \qquad (3.40)$$

Now, if the partial pressure of oxygen increases and accordingly the partial pressure of nitrogen decreases, keeping the pressure in the vessel constant at p_0, then the pressure exergy is different from environmental air. Such an air is called oxygen-rich air. In the present case, we have

$$n_1 + n_2 = n_{10} + n_{20} = n \text{ (total mol number)} \qquad (3.41)$$

$$x_1 + x_2 = 1, \quad x_{10} + x_{20} = 1. \qquad (3.42)$$

As an example, we shall find the exergy value of oxygen-rich air, where the oxygen is three times richer than atmospheric air, which has a molar ratio of 21(O_2)/79(N_2). The molar ratio of the new air is 63(O_2)/37(N_2), so that the x-values are $x_1 = 0.63$, $x_2 = 0.37$, $x_{10} = 0.21$, and $x_{20} = 0.79$. When these values are substituted into the Eq.(3.40), we have

$$E_n = pV[x_1 \ln(x_1/x_{10}) + x_2 \ln(x_2/x_{20})]$$

$$= 41.67 \text{ [kJ/m}^3\text{]} \qquad (3.43)$$

Equation (3.43) is obtained by substituting Eqs.(3.40) and (3.42) into the relationship:
$$E_n = E_1 + E_2. \qquad (3.43)$$

The equation of state for an ideal gas is also applied.

(b) Wet air. Equation (3.40) can also be applied to wet air having high humidity. If the subscripts 1, 2, and 0 show the water vapor, dry air, and the wet air in the environment, respectively, then we have, for a unit volume of air,

$$E_n = p[x_1 \ln(x_1/x_{10}) + (1 - x_1) \ln[(1 - x_1)/(1 - x_{10})]]. \qquad (3.44)$$

Equation (3.44) shows that the exergy of air drying is relatively large compared with refrigeration. For example, when air with a relative humidity of 70 % is dried to 40 %, an exergy of about 0.3 [kJ/m^3] is necessary. This value is about six times larger than the exergy needed to cool air from 303 [K] to 298 [K]. We should notice that the exergy of drying has a comparable or higher order of magnitude compared to refrigeration exergy.

(c) Fuel. Z. Rant has obtained a general evaluation formula for the chemical exergy of fuels. We shall cite his formulas.

(i) *Solid fuel.* The exergy of solid fuel (coal, wood, and so on) is

$$E = \text{LHV} - Lm \qquad (3.45)$$

where LHV, L and m represent the low heating value (p.16), the heat of evaporation for water, and the quantity of water contained in the solid fuels, respectively. This expression shows that the unavailable energy is due to water contained in the fuels.

(ii) *Liquid fuel.* The exergy of a liquid fuel such as gasoline, heavy oil, kerosene, light oil, and so on, is

$$E = 0.975 \text{ HHV}, \qquad (3.46)$$

where HHV is the higher heating value (p.16). The liquid fuel loses at least 0.25 % of its exergy when it burns. If the liquid fuel is stored in a vessel without a good airtight seal, then over a long time it will be partially oxidized and the exergy will be further reduced. The amount of reduction will be estimated by Eqs.(3.40) - (3.43).

(iii) *Gaseous fuel.* Rant's formula is

$$E = 0.95 \text{ HHV}, \qquad (3.47)$$

for the exergy available to gaseous fuel expressed by C_nH_m with n, m = 2, 3,···, *i.e.*, propane, propylene, butane, butyne, butylene, ethane, and methane.

The exergy values of some of those gases, at normal state, are shown in Table 3.1.

Table 3.1. Exergy of gaseous fuel (C_nH_m). Values are at 298.15 [K] and 1.01325×10^5 [Pa].

Name	Molecular formula	m/n	Exergy [kJ/mol]
propylene	C_3H_6	2	2,013.5
butane	C_4H_{10}	2.5	2,859.2
propane	C_3H_8	2.66	2,162.6
ethane	C_2H_6	3	1,503.0
methane	CH_4	4	834.8

Table 3.2. Exergy of some important chemical substances. The environment is taken to be saturated wet air at 298.15 [K] and 1.01325×10^5 [Pa] (after H. Kameyama et al[24]).

Name	Molecular formula(phase)	Exergy[kJ/mol]
carbon monoxide	CO (g)	279.67
carbon dioxide	CO_2 (g)	20.13
iron monoxide	FeO(s)	126.21
hydrogen	H_2 (g)	235.34
water	H_2O (l)	8.53
hydrochloric acid	HCl (g)	45.84
sulfuric acid	H_2SO_4 (l)	160.03
nitrogen	N_2 (g)	0.67
nitrogen dioxide	NO_2 (g)	55.63
salt	NaCl(s)	0
oxygen	O_2 (g)	3.93
sulfur dioxide	SO_2 (g)	310.86
hydrogen sulfide	H_2S(s)	809.0
silicon oxide	SiO_2 (s)	0

In this section, we have given only an outline of exergy analysis. Many kinds of exergies such as photon and electromagnetic phenomena, are not included. Moreover, most of the chemical substances such as hydrogen (H_2), methanol (CH_4O), carbon dioxide (CO_2), and others that play important roles in the production of chemical energy are shown in Table 3.2. However, the changes of exergy with chemical reaction are not introduced here, because they are rather practical and we have not enough space, to include them all.

3-2. Cost evaluation

(1) Primary energy

Primary energies are described in Chapter 1. They are classified as (i) fossil fuels (coal, oil, natural gas, *etc.*), (ii) nuclear energy, and (iii) natural energy (solar energy, hydro power, wind power, *etc.*). The cost of primary energy is composed of the searching cost (R), the development cost (D), and the operation cost (W). The total cost (T) is given by

$$T = R + D + W. \tag{3.49}$$

We shall consider, first, the searching cost. If the total cast paid to search for the primary energy resource is R_0, and the total quantity of the resource reserve is Q_0, then the depreciation to the search investment is given by

$$R = (R_0/Q_0)F(n) \tag{3.50}$$

where $F(n)$ is a function of the years n since the resource began producing. The unit of R is [$/t]. Even using modern, advanced technologies, it is very difficult and expensive to locate oil or gas fields. The chance of locating a big oil field with reserves of more than 10^8 [t] is less than 0.1 %.

The investment R_0 must be depreciated with an interest ϵ per year. If the life time of the resource reserves is n years and the depreciation is paid equally every year of production, then Eq.(3.50) is explained by

$$R = (R_0/Q_0) F(n) \tag{3.51}$$

with
$$F(n) = (1 + \epsilon)^n \epsilon /[(1 + \epsilon)^n - 1] \tag{3.52}$$

In the same way, we have for the depreciation of the development cost

$$D(n) = D_0 F(n)/Q_0 \tag{3.53}$$

The operation cost W is composed of (i) personal expenses, (ii) maintenance (repair, supply, *etc.*), (iii) earnings (dividend, saving, *etc.*, excluding interest), (iv) taxes, (v) insurance, (vi) preservation of safety, (vii) public associations, and (viii) other expenses.

The evaluation described above is the simplest way to estimate costs, and if the decrease of the reserved amount is taken into account, the factor Q_0 is replaced by $Q_0 G(n)$:

$$Q_0 G(n') = q \sum_{x=0}^{n'} (1-d)^x \qquad (3.54)$$

where q is the annual output quantity of the n'-th production year since the oil field or the coal mine or gas field first commenced. If d, the annual decline rate of the output, is independent of n and $q = q_0$ is a constant, then we have

$$Q_0 G(n) = (q_0/d)(1-d)^n \qquad (3.55)$$

Referring to Eqs.(3.50), (3.52), (3.53), (3.54), and (3.55), we may conclude that the low cost of a primary energy is obtained under the conditions : (i) large Q_0, i.e., the most important factor is that the total quantity of reserves is large, (ii) large n, i.e., the number of depreciation years is long, with small interest ϵ, (iii) small d, i.e., the annual decline rate of output must be small, and (v) small W.

However, the actual price of the primary energy is usually decided by the policies of the exporting and importing countries. The theoretical cost conditions are much better in the Middle East than in other regions.

(2) Secondary energy

As an example of secondary energy we shall use electrical energy. However, the discussion described below can be generalized to any secondary product, such as city gas, oil refining, and so on.

If the price of the secondary energy that is produced from a facility with an output capacity of N [kW] is T_s [$/kWh], then we have

$$T_s = [I F(n)]/(8,760w) + 860P/(\eta q) + W/(8,760N), \qquad (3.56)$$

where the first, the second, and the third terms of the right hand side of Eq.(3.56) represent the plant depreciation cost, the fuel cost, and the operation cost, respectively. The newly introduced parameters are: I [$/kW], the construction cost of the plant per [kW], 8,760w is the operation hours in a year; P the price of the primary energy per 1 [t] and 860 indicates the kcal equivalent to 1 [kW·h] (= 860 [kcal]); η, the conversion efficiency from heat energy (primary energy) to electrical energy (secondary energy), q[kcal/t], the heat energy density of the fuel; and W is the same as in Eq.(3.49).

Requiring a low price for T_s, it is obvious that a small I, W and large w, q and N are necessary. This will be commented on in the following discussion.

(a) **Scale merit.** We must take note of the last term containing W/N in Eq.(3.56), where the necessary operation is not always proportional to the output

Scale N. The ratio, in general, decreases with increasing plant size. Let the sum of the first and the third terms be T_{s0}, then we have

$$T_{s0} = AP^{-m} \quad (m > 0). \tag{3.57}$$

where P represents a plant size with output power N [kW], and m is an imperial numerical constant. Some examples for the value of the constant m are as follows: (i) for chemical plants, $m = 0.67$; (ii) for light water nuclear reactors, $m = 0.5$; and (iii) for oil thermal steam plants, $m = 0.7$.

Nevertheless, limitless large scale plants are not permitted because of the potential for heavy damage and disasters in case of emergency. The critical scale is decided by safety engineering considerations, case by case.

(b) Thermal electrical power plant. We shall take a recent example of an oil burning steam power plant in Japan. The predicted cost per [kW] of the investment for an oil burning plant, the annual interest, the annual operating rate, and the predicted years of operation (lifetime of the plant) are assumed to be 15 x10^4 [¥/kW], 6 [%/Y], 0.7 average over one year, and 20 [Y], respectively. Thus, we obtain 3.9 [¥/(kW·h)] for the first term of the right hand side of Eq.(3.56).

The price of oil per [t], the heat quantity of the oil per [t], and the conversion efficiency, are assumed to be 12.45 x 10^3[¥/t or 18$/B], 9.6 x 10^6[kcal/t], and 0.3, respectively, and the second term of the right hand side of Eq.(3.56) is 3.7[¥/(kW·h)].

Statistical data shows that the operation cost is about 3.5 - 4.5 [¥/(kW·h)], so that the cost price is about 11-12 [¥/(kW·h)]. On the other hand, the average electrical power rate in Japan is about 19 [¥/(kW·h)], the most expensive in the world. The difference between them, 7 - 8 [¥/(kW·h)], is considered to arise from the associated costs (investments and maintenance) of the power cables, transforming substations, pumping station dams, etc.

(c) Nuclear power plant. The predicted cost per kW of a light water nuclear power plant, the annual operation rate, and the price of nuclear fuel are assumed to be 32 x 10^4 [¥/(kW·h)], 0.8 average over one year, 27,000 [¥/kg or 240 $/kg], respectively. The other parameters are assumed to be the same as in the oil burning case, and so the first term of the right hand side of Eq.(3.56) becomes 7.3 [¥/(kW·h)], but the contribution from the fuel is estimated to be negligibly small (up to 1 [¥/(kW·h)]). Therefore, the contribution from the third term plays an important role in determining the electrical power price.

(d) Natural energy. The essential points of utilizing natural energies will be discussed by applying an extended formula of Eq.(3.56).

First of all, we notice that the term containing P becomes zero, and the investment cost I [$/kW] of a natural energy power plant must satisfy the following condition :

$$IF(n) \leq \eta gwE, \tag{3.58}$$

where E [kW·h/m^2] is the density of the incident natural energy averaged over one day per 1 [m²] of the plant, e.g., $E = 360$ [kW·h/m²] in the case of solar energy, g [$/kWh] is the price of the secondary energy, and w is the number of operative days ($w/8{,}760 = h/24$), h is the operating hours in a day of the plant in one year. Unless Eq.(3.58) is maintained, natural energy is not applicable in practice.

The utilization of wind power in windy areas and geothermal power plants are more efficient than solar energy because the number of operative days, w, is much greater.

(e) Solar energy. In the case of solar energy, we take I in units of [$/m²] as in Eq.(3.58), η is assumed to be 20 %, ϵ is 0.6 %, w is 2,000 [h], E is 1 [kW/m²]. Our problem is, "How many years are necessary to predicate the investment cost I in the case of a given price [$/(kW·h)] of electrical power?". If the cost of a solar energy plant is 500 [$/m²](this is the price for a thermal steam plant, a solar cell power plant being much more expensive) and the market price of electrical energy g is 0.1 [$/(kW·h)], then we have the answer of $n = 50$ [Y]. This is too long a lifetime for the plant to be realized, and the answer is "impossible" in practice. The reasons are, (i) I is too expensive, (ii) w is too small, (iii) E is too small, and (iv) g is too cheap. In other words, solar energy is of low density and intermittent in nature. A solar thermal steam power plant is thus comparatively difficult to construct unless some public policies are enforced.

An exceptional example of the practical application of solar energy utilization is the solar heater (water heater or room heater). We may take, in that case, $I = 300$ [$/m²], $g = 1 \times 10^{-4}$[$/kcal], and $\eta = 50$ %, then the repayment for the investment cost can be done in 4.1 [Y]. Numerous solar panel water heaters have been installed. Their number is now estimated at more than two million, with about half being operated in Japan.

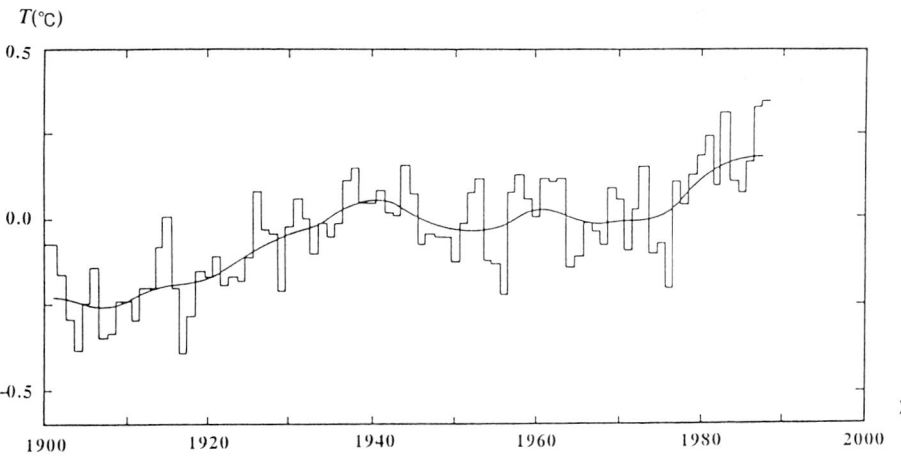

Fig. 3.3. Global surface air temperature. The standard level (0.0 °C) is the average from 1950 to 1979. (After University of East Anglia, U. K.)

(3) Political cost

(a) Global warming. The economic superiority of fossil fuels and nuclear energy have been demonstrated in the above section. For this reason, more than 88% of the primary energy in the world in recent years has been generated from fossil fuels. However, the exhaust gases from combusted fossil fuels have accumulated to an extent where serious damage is being done to the global environment. The exhaust gases can be classified into the three kinds and are shown in Table 3.3. The first group is composed of water vapor (H_2O), carbon dioxide (CO_2), and nitrogen dioxide (NO_2). These are not naturally injurious, and their emission control is impossible. Water vapor recycles in three days at most. Carbon dioxide is quite harmless, but contributes to the greenhouse effect (GHE) that locks heat within the atmosphere by preventing the escape of thermic rays from the earth. The accumulated amount is estimated at *ca.* 750×10^9 [cet] (carbon equivalent ton - one [cet] is equivalent to 11/3 [t] of CO_2), and the global temperature is increasing. The global warming trend from 1900 to 1990 is shown in Fig.3.3, where the century-long global warming is shown to be about 0.5°C.

Table 3.3. Gases exhausted from fossil fuel combustion

Group	Material	Climate effect	Injurious	Recycle time
1 (Difficult to exhaust control)	$H_2O(g)$	GHE:small humidity rise	no	3 days
	$CO_2(g)$	GHE:large	no	200 years
	$NO_2(g)(*)$	small	small	150 years
2 (Exhaust-controllable)	NOx	minute	direct acid-rain	no
	SOx	minute	direct acid-rain	no
	COx	minute	direct acid-rain	no
3 (Exhaust-controllable)	CnHm(**)	GHE:very large	indirect	no
	C	temperature drop	indirect	no
	Ashes	temperature drop	indirect	no

(*) NO_2 gas has less GHE (Greenhouse Effect), but N_2O gas has 290 times the effect compared to CO_2.

(**) CnHm has a large GHE, for example, CH_4 has 21 times as large an effect as CO_2.

Evaluation of Energy 109

We shall now mention the global damage due to the global temperature rise. Firstly, by unusual change of weather affects agricultural products and, secondly the surface level of the sea has risen. Assuming that the average depth of the seas in the world is 3,795 [m], the coefficient of volume expansion is $2 \times 10^{-4}/°C$, and the sectional areas of the seas are constant, then we can assume an approximate sea level rise of about 80 [cm/°C]. In addition, the ice mountains on the Antarctic continent will melt and the surface level of the sea will rise up an additional 30 - 40 [cm/°C]. Thus most of the coastal industrial and metropolitan zones will sink down below the seawater. The resulting damage is too huge to estimate.

The origins of global warming are energy utilization (57 %), CFCs (flons) (17 %), agriculture (14 %), and land development (9 %). Energy utilization implies the emission of exhaust gas which is almost entirely CO_2.

The second and third groups listed in Table 3.3 have both directly and indirectly harmful effects (e.g. disease of the respiratory organs and acid rain). However, these groups are subject to control in industrial countries, but are not yet controlled in developing countries.

It is expected that advanced technology will be more widely available to all countries in the near future, so that gas belonging to the second and third groups will no longer appear. The biggest problem then is how to control CO_2.

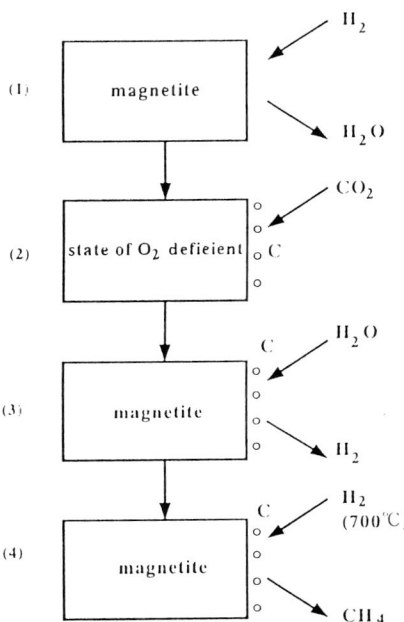

Fig. 3.4. Preliminary experiment of CO_2-fixation. By using the surface of magnetite (Fe_3O_4), CO_2 gas reacts on H_2 to produce methane (CH_4).

(a) Short term policy. In order to decrease the quantity of exhaust CO_2, some urgent undertakings must be initiated. Most of them require governmental or provisional policies. We shall list some of these below.

(i) *Development of new technologies for synthesizing useful substances utilizing CO_2*, for example, the catalytic hydrogenation method in the presence of the catalyzer, $Co/ZnO/Al_2O_3$, CO_2 and H_2 used to produce metanol (CH_4O). This method needs gas pressures as high as 10^5 [hPa]. To generate this high pressure, electrical energy is needed, which is first converted from a primary energy source. Therefore such a chemical technology does not always effectively reduce CO_2. New technology developments that need less energy are necessary.

Another example is shown in Fig.3.4. Prof. H. Tamaura (Tokyo Institute of Technology) has developed a new technology to fix CO_2 and manufacture methane. (1) Hydrogen gas (H_2) is blown against the surface of magnetite (Fe_3O_4). (2) Water vapor (H_2O) is generated and the surface becomes deficient in oxygen (O_2). (3) Next, CO_2 is blown on the surface; the oxygen is absorbed by the surface yielding carbon. (4) Hydrogen gas at high temperature (700°C) is blown against the surface and methane (CH_4) is produced. This system is scientifically valid, but is difficult in practice, since the surface degrades when the reaction is repeated and CO_2 is emitted in process (3). However, such innovative research is highly recommended.

(ii) *Development of geoengineering.* Geoengineering occurs on a large scale and is expensive, but it is most desirable. Carbon in the deep sea below depths of 3,700 [m] is dissolved as bicarbonate ions (HCO_3^-). Therefore it is possible to utilize the reaction:

$$CaCO_3 + CO_2 + H_2O \rightarrow Ca^{2+} + 2HCO_3^-, \qquad (3.59)$$

since calcium carbonate ($CaCO_3$) exists plentifully in the deep sea bed.

Other methods have been proposed to confine CO_2 in the deep sea, *e.g.*, to compress CO_2 to a pressure greater than 1.4×10^5 [hPa] to obtain liquified CO_2 which is stable in the deep sea at a depth of more than 3,000 [m]. The sea water at this depth has a density as high as that of the liquid CO_2 so that stability is realized. However, this method requires much energy.

(iii) *Development of alternative energy.* As discussed in the previous sections, the utilization of natural energy can hardly be competitive economically at present. Therefore, some promotion policies are needed to encourage the development and spread of these technologies and their practical applications.

Concrete measures for encouraging this development are as follows.
(1) Subsidies for the interest ϵ in Eq.(3.52). (2) Purchasing the energy generated by natural power plant from a **domestic plant** when it is in excess. (3) Loans to the plant. (4) Deduction of the necessary expenses for the domestic plant (such as solar cell or windpower power plants) from individual income tax. (5) Others.

Many industrialized countries such as the U.S.A., Germany, Netherlands, Japan, and so on have developed systems such as (1), (2), and (3) above. Excess electrical energy can be purchased by the electric power company at a lower price than that of which the company sells. This system must be considered and the price difference should be supplied by public expense. Measure (4) above is the most desirable although it has not yet been undertaken.

There exists only one parameter that can be a policy object in Eq.(3.58), *i.e.*, the interest e. Hence it is not easy to design effective policy for promoting natural energy development.

(**iv**) *Nuclear power plants.* Nuclear power plants do not necessarily emit GHE gases. Although the construction and the operation of a plant requires the application of fossil fuels that emit CO_2, the amount is estimated to be below 10 % compared with fossil fuel combustion. Nuclear energy is definitely to be preferred, but the reserves of nuclear fuels are not in excess of those oil, and the radioactive materials exhausted from the nuclear plants have another damaging effect upon the environment. Moreover, safety management requires considerable expense, in order to ensure system security.

(b) Long term policy.

(i) *Mitigation and adaptation.* The development of technology to counter CO_2 problem is an important short- and long-term policy issue. Another policy solution is, economical, and can be divided into short-term and long-term policies, the former being called "mitigation" which copes with the problem in advance, and the latter being called "adaptation". Adaptation is undertaken after the environmental and ecological consequences of climate change occur. For example, enormous expense is necessary to carry out counter-measures against the sinking of coastal cities and industrial zones under the sea.

The question of whether mitigation or adaptation is more advantageous depends on the nature of policy and the time scale of the undertaking. In order to carry out the same economic policy, mitigation is in general more profitable, that is, the global real income is more than that for adaptation. For example, $1,000 invested at the present time is equivalent to $1,060 after one year. If no mitigation is undertaken and 5×10^{12} dollars will be needed to carry out adaptation counterplans after 100 years, then only 148×10^8 dollars are necessary to undertake the same counterplans right now.

Figure 3.5 (after a report[32] published by National Academy Press, 1991) shows four cases. (1) A case without climate change and no policy. The real income grows and the growth is plotted by a predicted trend taken from past history. (2) Mitigation is undertaken, before adaptation becomes necessary. This case seems st, because climate change is now actually occuring. (3) No mitigation is done so that adaptation will be occasionally undertaken. The real income is reduced, next to case 4. (4) No mitigation and no adaptation lead to this case. Climate change results in a large amount of damage which will require a very large investment to remedy, so that the real income will be smallest.

We can now see that mitigation is recommended qualitatively. The next section will deal with the quantitative analysis.

112   Energy Technology

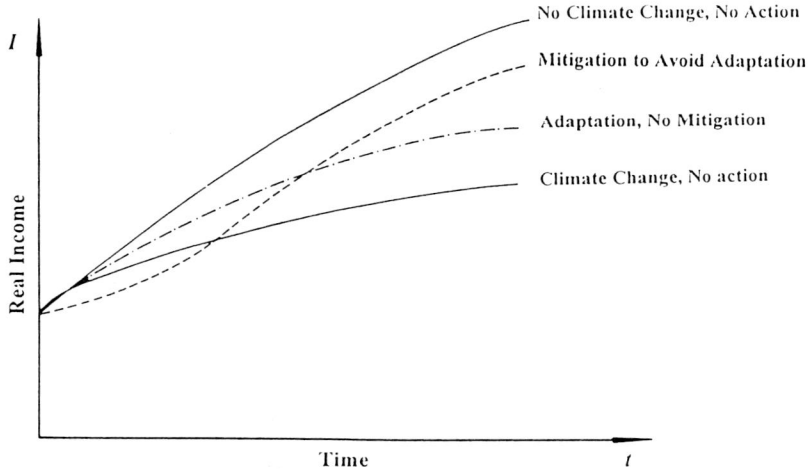

Fig. 3.5. Schematic comparison of mitigation and adaptation (after National Academy of Science, National Academy of Engineering, Institute of Medicine, U. S. A., 1991).

(ii) *Nordhaus' theory*[35]. A concrete theory of mitigation against climate change is a conspicuous problem in economics. Prof. W.D. Nordhaus (Yale University, U.S.A.) published a reliable theory in 1992. His analysis is achieved by a specialized application of nonlinear optimization and hence a detailed picture is not introduced here. His method is outlined very simply in the following, but the results obtained are important and are shown in Figs. 3.6 - 3.10.

His treatment is after a method called the DICE model (dynamic integrated climate-economy) which is a dynamic representation of the economical impact of the cost policies for controlling GHG (greenhouse gas) emission.

GHG includes gases other than CO_2 but because CO_2 occupies more than 80 % of the emissions it is the only object of the analysis. Controlling GHG emission is carried out by imposing additional taxes on fossil fuels or a newly introduced tax system according to the quantity of CO_2 emission.

The fractional loss of global output from GHE warming is assumed to depend on the temperature increase.

Raising the price of fossil fuels results in a reduction of economic growth by reducing output, however if no such mitigation is undertaken, then a huge amount of expense is required to establish counterplans against the climate change over a long time. That another optimal path exists is the subject of Nordhaus' study. Optimization is done by maximizing an objective function that is the sum of the utilities of capital consumption:

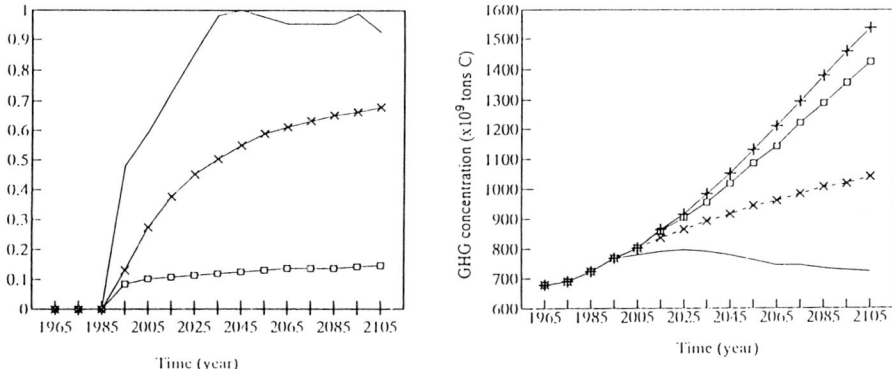

Fig. 3.6. Greenhouse gas control rate. The control rate is shown for the optimal case (□), for the uncontrolled case (+), the emission stabilization case (x), and the case of the limiting temperature change to 1.5 °C (-).

Fig. 3.7. Greenhouse gas concentrations. Concentrations of green-house gas (CO_2 and CFCs) are shown for the uncontrolled case (+), the optimal case (□), the emission stabilization case (x), and the climate stabilization case (-), respectively (after W. D. Nordhaus, 1992).

$$\text{Max} \atop c(t) \sum_{t=0}^{\tau} \frac{P(t)c(t)}{(1+\rho)^t}, \qquad (3.60)$$

where $P(t)$ and $c(t)$ are the population and the flow of consumption per capita, respectively, at time t. Parameter ρ is called a rate of time preference meaning that the longer the time the larger is the time effect (if $\rho = 0$, there exists no direct time effect).

Maximization of Eq.(3.60) is carried out under some conditions that are rather complicated and depend on historical statistics. The first set of constraints relate to the growth of output, and the second set are used to link the GHG emission → climate change → economical outgoings.

The total output $Q(t)$ is composed of the consumption part $c(t)$ and the investment part $I(t)$ and can be expressed by

$$Q(t) = X(t)A(t)K(t)^{\gamma}P(t)^{1-\gamma}, \qquad (3.61)$$

where $X(t)$ indicates the impact due to climate change, $A(t)$ is a term expressing technology development and application, $K(t)$ is capital, and P is labor power. The

114   Energy Technology

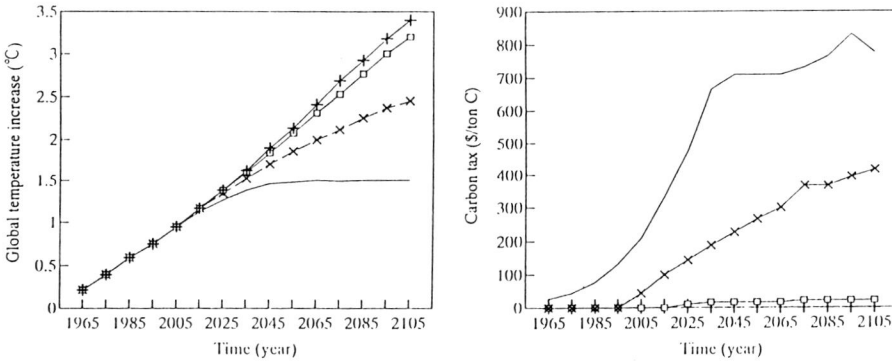

Fig. 3.8. Projected global mean temperature. The temperatures are shown for uncontrolled case (+), the emission stabilization case (x), and the climate stabilization case (-), respectively (after W.D. Nordhaus, (1992).

Fig. 3.9. Carbon taxes in different policies. The carbon taxes calculatedare shown for the optimal case (□), the emissions stabilization case (x), and the climate stabilization case (-), respectively (after W. D. Nordhaus, 1992).

parameter γ is the elasticity of output with respect to capital.

Figure 3.6 is plotted after the DICE model substitution, using the historical data and limiting the temperature change to 1.5°C at the most.

Figures 3.7 and 3.8 are plotted following scientific and economic analyses that are too complicated to be presented here. However, we shall show some interesting formulas that are useful to envision Nordhaus' treatise.

The first is the relationship between CO_2 emission $E(t)$ and the output $Q(t)$:

$$E(t) = [1 - \mu(t)]\sigma(t)Q(t), \qquad (3.62)$$

where $\mu(t)$ is the emission control rate and a coefficient $\sigma(t)$ is decided by historical data.

The second is the relationship between CO_2 concentration relative to the pre-industrial time $M(t)$ and the CO_2 emission:

$$M(t) - 0.9917M(t-1)$$

$$= 0.64(\pm 0.15)E(t). \qquad (3.63)$$

Third is the relationship between global temperature increase $\Delta T(t)$ and the

Evaluation of Energy 115

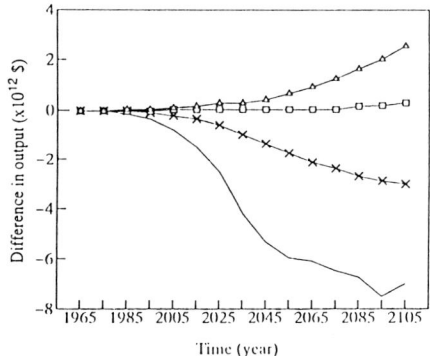

Fig. 3.10. Impact of policies on global output. The calculated difference in economic welfare between a policy and a noncontrol approach is shown for the optimal policy case (□), the geoengineering case (Δ), the emission stabilization case (x), and the climate stabilization case (-), respectively (after W. D. Nordhaus, 1992).

fractional loss $d(t)$ of global output due to GHE warming:

$$d(t) = 0.00148[\Delta T(t)]^2. \tag{3.64}$$

The fourth is the relationship between the emission control rate $m(t)$, temperature rise $\Delta T(t)$, and the climate impact factor $X(t)$ in Eq.(3.61).

$$X(t) = [1 - 0.0686\, \mu^{2.889}(t)] / [1 + 0.00148\{\Delta T(t)\}^2]. \tag{3.65}$$

Equations (3.61) through (3.65) are not enough to plot Figs.3.7 through 3.10, but the trends of the curves can be understood.

We must pay attention especially to Figs. 3.9 and 3.10 where climate stabilization requires too much tax expenditure. The outgoings are also too much. These trends suggest that climate stabilization is nearly impossible. Although the carbon tax as well as the output for adaptation in the future is very small, the temperature rise will reach 2.2°C in 2060 when oil and gas will be exhausted. A global temperature rise of 2.2°C is considered to be destructive. Therefore much more effective geoengineering technologies must be developed as soon as possible.

Prefixes

exa	E	10^{18}	deci	d	10^{-1}
peta	P	10^{15}	centi	c	10^{-2}
tera	T	10^{12}	milli	m	10^{-3}
giga	G	10^{9}	micro	m	10^{-6}
mega	M	10^{6}	nano	n	10^{-9}
kilo	k	10^{3}	pico	p	10^{-12}
hecto	h	10^{2}	femto	f	10^{-15}
deca	da	10	atto	a	10^{-18}

Chapter 4
Energy Systems

LNG tanker

Chap. 4. Energy Systems

4.1. Energy Transfer System

(a) Energy system. The term "energy system" is ambiguously used to depict a total system combining elemental subsystems, such as the search for primary energy resources, and its subsequent development, refining, conversion, transportation, storage, distribution, utilization, security, pollution problems, and so on. Accordingly, the energy system has been regarded as a kind of economic system attaching importance to the distribution system rather than an object of science and technology.

On the other hand, the term "system" in engineering fields represents a whole system where each elemental subsystem is organized in close cooperation with each other. System engineering studies how to optimize the output of the system function by the selection and combination of parts and technologies. Such system engineering is the main trend in modern technology rather than the innovation of elemental hardware. Examples include cameras, audio-visual equipments, *etc.*, in daily life, and the bullet train, electrical power line, *etc.*, as public services. The importance of system engineering is mostly emphasized in big national projects such as the Apollo project (U.S.A.), strategy, *etc*.

Energy is the lifeblood of civilization and hence the energy system may be compared to the network of blood vessels in a human body. One may understand the important role of the energy system by using this metaphor, but it is regrettable that there exists no established scientific field covering the energy system. We have many innovative and advanced energy conversion technologies such as those described in this book. Once these technologies are put in order and systematized following basic ideas then an optimal energy system is obtainable. The present chapter is intended to be an introduction to such an approach.

An outline of energy flow is shown as a system in Fig.4.1. Primary energy (fossil fuels, nuclear energy, renewable energy) is transported, stored, and converted to secondary energy (electrical power, city gas, gasoline, light oil, kerosene, *etc*. The secondary energy is transported (or transmitted), stored and converted to a more convenient energy with high quality. By quality of energy we mean, for example, electrical energy with constant voltage and frequency, city gas with constant pressure and caloric density, gasoline with high octane and no impurity. The term "tertiary energy system" is applied to the most convenient utilization system.

The energy flow is accompanied by unavailable and dispersed energy which can be referred to as "losses". We shall comment on such an actual energy system later.

The actual energy system is simplified and shown in Fig.4.2 as a flow diagram where the sign →| denotes the end of flow. Primary energy can be converted to plural secondary energies, *e.g.*, co-generation systems, solar energy conversion to both electrical power and heat. Such multistage energy utilization is a

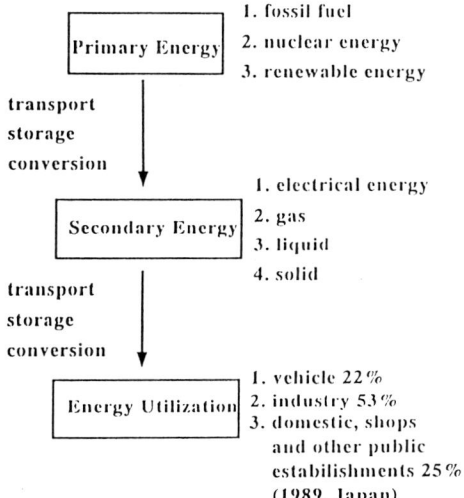

Fig. 4.1. Energy system. Another typical utilization is shown in Fig.4.6.

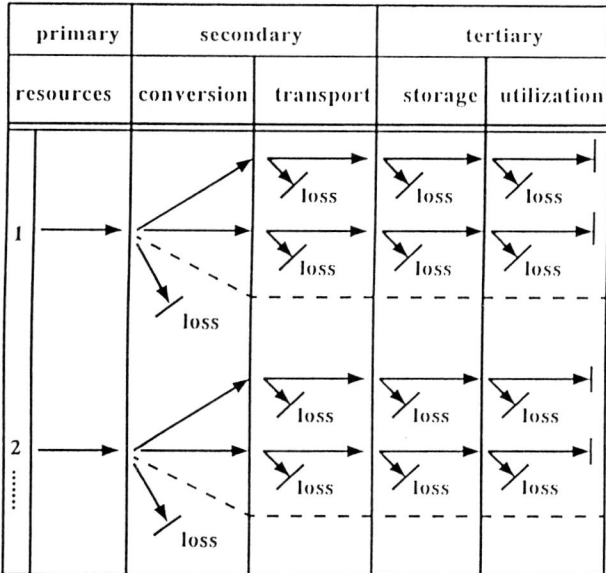

Fig. 4.2. Energy transfer system. The primary, secondary, and tertiary energy systems are energy resource, energy carrier, and utilization system, respectively. Transport and storage are common to both the secondary and tertiary system.

noticeable trend in the energy saving age, and can be a primary object of energy system science.

The losses in transport and storage are unavoidable. For example, oil or gas transportation by a pipeline system needs energy to compress the fluid, which is one kind of loss.

Figure 4.2 does not include the case where the primary energy generates secondary energy of almost the same quality. Such a system actually exists, for example, the electric power utility line system of 110 volts and 50 Hz cycle, connected with solar cell power generators in Japan.

(b) Transfer matrix and efficiency vector.[2] An energy transfer system is defined by the flow diagram shown in Fig.4.2, the component subsystems being the primary, secondary, and tertiary.

The elements of each subsystem have two functions. The first is its transferring role, *i.e.*, from 1 to 2, and from 2 to 3. The second is its terminating role as shown by →|.

The quantity of the i-th primary energy is denoted by P_i and its input to the i-th secondary energy S_i, thus we have a vector of primary system

$$P = (P_1, P_2, \cdots P_i, \cdots P_n). \tag{4.1}$$

The secondary and tertiary systems can be expressed by a single matrix because the secondary system is determined by the terminal utilization (tertiary system).

Let us take a secondary system for k energies whose subsystems are

$$S_1, S_2, \cdots S_k,$$

and these are transferred to a tertiary system whose subsystems are

$$S_{k+1}, S_{k+2}, \cdots, S_{k+l} \quad \text{with } n \geq k+l$$

The matrix element t_{ij} ($1 \leq i, j \leq n$) expresses an energy transfer between the subsystems and is defined by the "transfer rate of energy" from S_i to S_j. This transfer rate is a similar quantity expressed by Eq.(2.31). Strictly speaking, λ_m of Eq.(2.31) does not include dispersed energy, while t_{ij} includes every loss.

We assume, for the secondary system,

$$t_{ij} = 0 \quad (j \leq k) \tag{4.2}$$

and, for the tertiary system,

$$t_{ij} = \delta_{ij} \quad (k+1 \leq j \leq n) \tag{4.3}$$

The transfer matrix is thus defined as :

$$T = \begin{bmatrix} 0 & t_{12} & t_{13} & \cdots & t_{1k} & t_{1k+1} & t_{1k+2} & \cdots & \cdots & t_{1n} \\ t_{21} & 0 & t_{23} & \cdots & t_{2k} & t_{2k+1} & . & \cdots & \cdots & t_{2n} \\ t_{31} & . & & : & t_{3k+1} & & . & & : \\ : & & . & t_{k-1k} & : & & & . & t_{kn} \\ t_{k1} & t_{k2} & \cdots & t_{kk-1} & 0 & t_{kk+1} & t_{kk+2} & \cdots & \cdots & t_{kn} \\ \hline 0 & 0 & 0 & \cdots & 0 & 1 & 0 & 0 & \cdots & 0 \\ 0 & 0 & & \cdots & \cdots & 0 & 0 & 1 & \cdots & \cdots & 0 \\ 0 & & . & & : & 0 & & . & & : \\ : & & & . & 0 & : & & & . & 0 \\ 0 & \cdots & \cdots & \cdots & 0 & 0 & \cdots & \cdots & \cdots & 1 \end{bmatrix} \quad (4.4)$$

Now, we shall define the efficiency vector. If the primary vector (4.1) is given, then the energy quantity W_f that is transferred to S_f of the tertiary system is written as

$$W_f = \sum_{i=0}^{n} P_i e_{if}, \quad (4.5)$$

where e_{if} means the ratio of the quantity that S_i obtained from P_i to that transferred to S_f. The efficiency vector is thus defined by

$$e_f = \begin{bmatrix} e_{1f} \\ e_{2f} \\ . \\ . \\ e_{nf} \end{bmatrix} \quad (4.6)$$

Then it is obvious that

$$W_f = P \cdot e_f. \quad (4.7)$$

Energy is transferred from S_i ($1 \le i \le k$) to S_j($k+1 \le j \le n$) by the transfer rate t_{ij} and from S_j to S_k by the efficiency e_{jk}, therefore the overall efficiency is expressed by

$$t_{ij} e_{jk}.$$

The system efficiency (overall efficiency) is given by

$$e_{if} = \sum_{j=1}^{n} t_{ij} e_{jf}. \quad (4.8)$$

122   Energy Technology

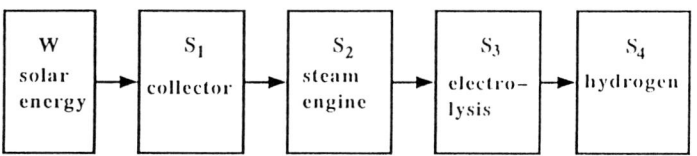

Fig. 4.3. Hydrogen production system using solar energy.

with $e_{ff} = 1$, $e_{if} = 0$ ($i \neq f$, $k+1 \leq i \leq n$)
or
$$e_f = T\, e_f. \tag{4.9}$$

A similar expression for efficiency introduced in the above can be generalized to an actual system such as the case where the energy circulation makes a loop in the secondary subsystem.

4-2. Examples of Energy Systems

(1) Simplest example

The simplest example of an energy system is a hydrogen production system by electrolysis using the electrical energy generated by a steam engine whose heat source is given by collected solar energy (Fig.4.3).

In the present case, the transfer matrix is

$$T = \begin{bmatrix} 0 & t_{12} & 0 & 0 \\ 0 & 0 & t_{23} & 0 \\ 0 & 0 & 0 & t_{34} \\ 0 & 0 & 0 & 1 \end{bmatrix} \tag{4.10}$$

where, $t_{12} = \eta_c$ (collector efficiency), $t_{23} = \eta_{te}$ (efficiency of steam engine), $t_{34} = \eta_{el}$ (electrolysis efficiency). The efficiency vector is written as

$$e_4 = T\, e_4$$

(4.11)

which gives

$$\begin{Bmatrix} e_{14} \\ e_{24} \\ e_{34} \\ e_{44} \end{Bmatrix} = \begin{Bmatrix} 0 & \eta_c & 0 & 0 \\ 0 & 0 & \eta_{te} & 0 \\ 0 & 0 & 0 & \eta_{el} \\ 0 & 0 & 0 & 1 \end{Bmatrix} \begin{Bmatrix} e_{14} \\ e_{24} \\ e_{34} \\ e_{44} \end{Bmatrix} \tag{4.12}$$

with $e_{44} = 1$.
The input vector P is

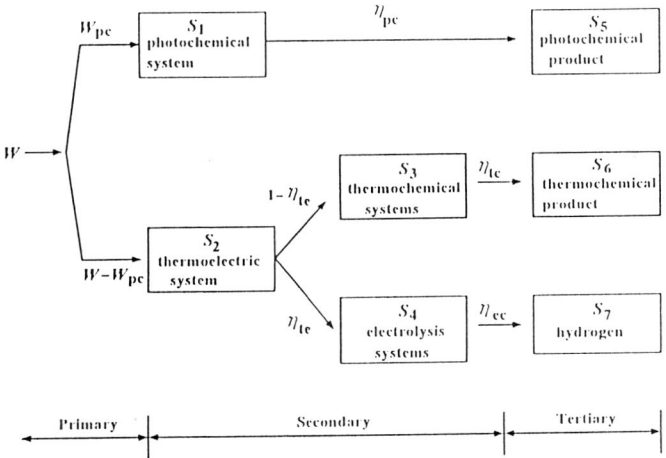

Fig. 4.4. A multistage utilization of solar energy. Photochemical conversion and thermoelectric conversion are simultaneously applied. Unavailable heat is applied to advance the thermochemical system. Generated electrical power is applied to water electrolysis.

$$P = (W, 0, 0, 0)$$

Then we have, for the system efficiency,

$$\eta_0 = (P/P) \, e_4 = \eta_c \eta_{te} \eta_{el}. \tag{4.13}$$

(2) Solar-hydrogen energy system

Ohta's group at the Yokohama National University[4] has studied a multistage solar energy utilization system for splitting water. The resulting hydrogen is a clean and powerful energy carrier.

The overall system is shown in Fig. 4.4. First solar energy with a short wavelength W_{pc} is used to promote a photochemical reaction such as illustrated in Eqs.(2.95)-(2.100). The remaining solar energy $(W - W_{pc})$ is collected to heat the p-n junction of a semiconductor thermocouple. This thermoelectric converter generates a large amount of unavailable heat as described on page 62 so that the waste heat is utilized to advance a thermochemical reaction (a concrete example will be introduced on page 202).

The primary, the secondary, and the tertiary systems are shown in Fig.4.4. Considering that S_1, S_2, S_3, S_4 are the secondary system, and S_5, S_6, S_7 are the tertiary system, we have

$$T = \begin{bmatrix} 0 & 0 & 0 & 0 & \eta_{pc} & 0 & 0 \\ 0 & 0 & 1-\eta_{te} & \eta_{te} & 0 & 0 & 0 \\ 0 & 0 & 0 & 0 & 0 & \eta_{te} & 0 \\ 0 & 0 & 0 & 0 & 0 & 0 & \eta_{ec} \\ 0 & 0 & 0 & 0 & 0 & 0 & 0 \\ 0 & 0 & 0 & 0 & 0 & 0 & 0 \\ 0 & 0 & 0 & 0 & 0 & 0 & 0 \end{bmatrix} \quad (4.14)$$

and the efficiency vectors are

$$e_5 = \begin{bmatrix} \eta_{pc} \\ 0 \\ 0 \\ 0 \\ 1 \\ 0 \\ 0 \end{bmatrix}, \quad e_6 = \begin{bmatrix} 0 \\ (1-\eta_{te})\eta_{te} \\ \eta_{te} \\ 0 \\ 0 \\ 1 \\ 0 \end{bmatrix}, \quad e_7 = \begin{bmatrix} 0 \\ \eta_{te}\eta_{ec} \\ 0 \\ \eta_{ec} \\ 0 \\ 0 \\ 1 \end{bmatrix} \quad (4.15)$$

The input vector is

$$P = (W_{pc}, W - W_{pc}, 0, 0, 0, 0, 0) \quad (4.16)$$

from which the following energies are given to each subsystem of the tertiary system:

W_c(photon energy) $= W_p = Pe_5 = W_{pc}\eta_c$ to S_5,

W_q (thermochemical energy) $= Pe_6 = (W - W_{pc})(1 - \eta_{te})\eta_{tc}$ to S_6,

W_E(electrochemical energy) $= Pe_7$

$= (W - W_{pc})\eta_{te}\eta_{ec}$ to S_7,

where η_c, η_{te}, η_{tc}, and η_{ec} represent photochemical, thermoelectric, thermochemical, and electrochemical efficiencies, respectively.

The overall system efficiency is given by

$$\eta_0 = (W_c + W_q + W_e)/P$$

$$= [W_{pc}\eta_{pc} + (W - W_{pc})(1 - \eta_{te})\eta_{tc}$$

$$+ (W - W_{pc})\, \eta_{te}\, \eta_{ec}\,]/W. \qquad (4.17)$$

(3) Multistage conversion

The effective utilization of energy resources is one of the most important tasks for energy engineering, because it provides for the conservation of resources and the reduction of global environmental problems. A useful secondary energy is obtained by an energy conversion process from a resource, but it is accompanied by unavailable and dispersed energies. This system is shown for the five kinds of energies in Table 4.1.

Unavailable energy is usually able to generate a useful secondary energy, while dispersed energy is not. For this purpose, multistage energy conversion technology has been developed.

The system of multistage energy conversion shown in Fig.4.3 uses a successive conversion of secondary energies to obtain upper grade quality energy, but the overall efficiency is low. Conversely, the present object is to utilize effectively the unavailable (wasted) energy so that the overall successive conversions makes the efficiency very high (Fig.4.5). Although the quality is not upgraded, the efficiency is given by

$$e_0 = 1 - \prod_{i=1}^{n} (1 - e_i), \qquad (4.18)$$

Table 4.1. Energy conversion and unavailable energy

Energy	Resource	Conversion	Unavailable	Dispersed
mechanical	hydro power wind power	turbine propeller	slower-motion	friction viscosity
electromagnetic	no available resources	generator	recovered	joulian heat discharge
photon (particle beam)	solar beam (nuclear energy)	solar cell, photochemical reaction	longer (shorter) wavelength	heat
chemical	fossil fuel density gradient	electrolysis, chemical reaction	lower temperature, chemical product (low grade)	heat, unneeded-product
heat	nuclear fission, fossil fuel, wood, geothermal, solar, ocean thermal	combustion collector, heat pump	lower-temperature	near the enviromental temperature

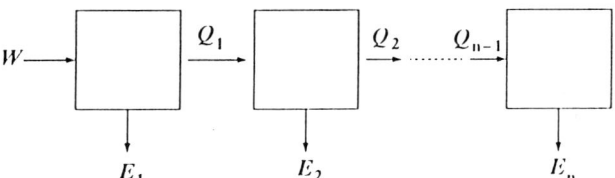

Fig. 4.5. Multistage energy conversion in series. Unavailable energies are successively converted.

where e_i is the efficiency in the i-th conversion. An analogous discussion is possible using exergy analysis. Some examples follow.

(a) A typical example is a co-generation system. A co-generation system utilizes the exhausted heat from a Diesel electric power generator. The exhausted heat (unavailable energy from the first energy converter) is available to generate steam or hot water that is used for room warming.

Electric power generation by a Diesel engine is 36 %, and the available rate of conversion from waste heat to a steam or hot water generator is 50 %, then the overall efficiency is

$$e_0 = 1 - (1 - 0.36)(1 - 0.5)$$

$$= 0.63$$

(b) IGCC is an abbreviation for "integrated gas combined cycle". This is a proposed system for effectively utilizing coal gasification. Gasified coal is combusted to evolve heat at a temperature higher than 1,000°C, which is then applied to drive a gas turbine. The wasted heat, with temperature higher than 400°C, is used to generate water vapor, which is applied to drive a vapor turbine. If the efficiency of the former and the latter are 48 % and 34 %, respectively, the overall system has an efficiency of

$$e_0 = 1 - (1 - 0.48)(1 - 0.34)$$

$$= 0.66.$$

The Cool Water station of Edison Electric Power Co. in Southern California has a capacity of 1.5×10^5 [kW] of IGCC, combusting 1,000 [t] of coal per a day since 1984. The IGCC in Japan is a system utilizing LNG (liquified natural gas) effectively. The total capacity was ca. 4×10^6 [kW] in 1991.

(4) Actual system

About 8.4×10^9 [oet] (oil equivalent ton = 12.2×10^6 [kcal]) of energy are consumed each year in the world. For developed countries, the final consumption

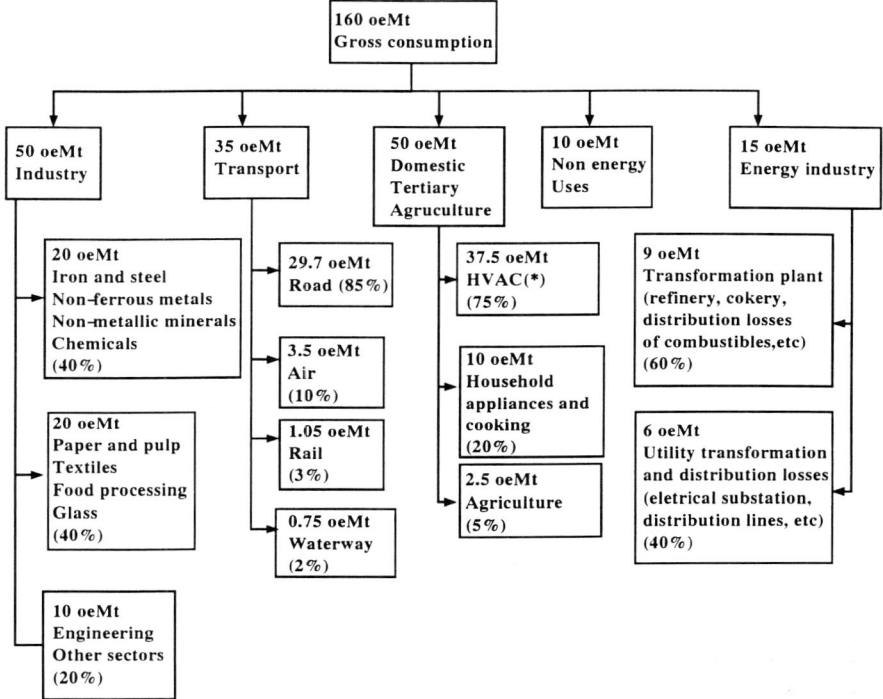

Fig. 4.6. Final consumption of energy in a typical industrial country. oeMt = oil equivalent million ton. Japanese data is shown in Fig.4.1. (*)HVAC means heating, ventilation, air conditioning. (after G. Petrecca : *Industrial Energy Management : Principles and Applications*, 1993, Kluwer Academic Publishers).

is distributed roughly as 31 % for industry, 22 % for transport, 31 % for residential, tertiary, and agriculture, 6.5 % for non-energy uses (manufacturing of plastics *et al.*), and 9.5 % for the energy industry. Such a typical sharing is shown in Fig.4.6[40]. The above percentage for transport is not valid in North America where the share is more than 35 %. Energy for transport is almost exclusively gasoline or light oil for automobiles in Europe and North America, but the use of electric trains is appreciable in Japan.

Some comments regarding Fig.4.6 are described below.

(a) Primary energy is selected in expectation of its terminal utilization. In other words, the most adequate primary energy is chosen and supplied to the final consumption system. For example, coal is for iron and steel manufacturing, heavy oil is for ships (waterway), hydrogen is for rockets, solar energy is for domestic heating and air conditioning, and wind and hydro power are for agriculture.

(b) The utilization of electrical energy is recommended to be limited in *ad hoc* fields. As has been discussed in the section on exergy, it is wasteful to heat up room by electrical power Joule heating. Instead the heat collected by solar collectors is preferred.

Electrical power should be applied to communication, TV, radio, fax, telephone, train, motor, and other applications exclusively driven by electrical energy. Electrical and electronic machines have high efficiency and low unavailable energy. Electrical energy is precious because it is generated and transmitted by many processes. The electrical energy of 1 [kW·h] has a caloric equivalent of 860 [kcal], but to generate 1 [kW·h] of electrical energy by a heat engine requires more than 2,530 [kcal] input heat.

(c) The cost of secondary energy can be determined by the balance of investment costs, fuels, and other systems for safety, environments, *etc.* The key for solving the question "*How to optimize the cost*" is undoubtedly the development of new technology, which is along the traditional trend rather than by new innovations. Good examples are: (1) Power generation by gas turbine on ingenious cooling system and with blades that can endure a temperature high enough to give a Carnot efficiency as high as 50 %. Such a turbine system is preferred to MHD power generation (ref. p.74). (2) SPE (solid polymer electrolysis) and VPE (vapor phase electrolysis) have been studied and restrain the development of thermochemical water splitting (ref. p.128). SPE uses Nafion film and electrolyzes water vapor as high as 1,000°C and VPE uses Zr-Y or $LaMnO_3$-Sr alloy to get hydrogen from water vapor at 1,300°C.

(d) **Hybrid system.** As one of the oil-saving contrivances, COM (coal oil mixture) fuel has been invented and utilized since the beginning of the 1980s. If M [kg] of COM is composed of M_1 [kg] of heavy oil and M_2 [kg] of powdered coal, the energy density Q is

$$Q = [M_1 Q_1 + M_2 Q_2] / M, \tag{4.19}$$

where Q_1 and Q_2 are the energy density of heavy oil and coal, respectively.

A similar technology called CWM (coal water mixture) has been studied and implemented. When powdered coal is combusted in the presence of enough air, the temperature is high enough so that NOx is exhausted. To reduce NOx, sprayed CWM is burned. The heat of vaporization of water (= 2,444 [kJ/kg]) absorbs the heat and the temperature goes down. In addition, NOx is dissolved into the vapor so that the exhausted NOx is appreciably reduced.

The combustion temperature of CWM is

$$T = T_0 (31{,}250\, M_1 - 2{,}444\, M_2) / 31{,}250\, (M_1 + M_2), \tag{4.20}$$

where 31,250 [kJ/kg] is the energy density of coal, M_1 and M_2 are the weight of coal and water, respectively, and T_0 is the temperature generated by the coal combustion.

(5) Hydrogen energy system

A present system for electrical power storage is shown in Fig.4.7, where $P_1, P_2, ...$ show energy resources such as fossil fuels, hydropower, nuclear power

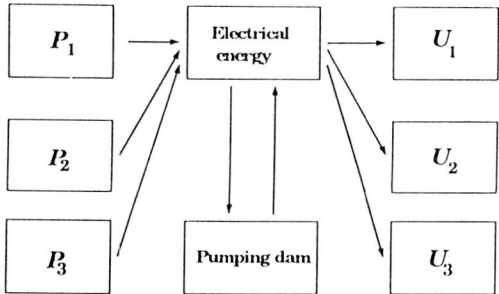

Fig. 4.7. The present system for electrical power storage. P_is are resources and U_is are utilization systems.

etc. When electrical energy is in excess, fuel combustion is stopped and the water falling down from the dam is stopped, that is, they are stored as fuels and potential energy, respectively. Surplus energy is also used to pump up water to a dam as high as several hundred meters.

In the industrial countries, electrical energy needs fluctuate up to three times a day. For example, during a mid-summer afternoon in Japan electric power needs for refrigeration are about three times that compared to the morning of January 2nd when all factories, trains, schools, *etc.* are not working and street lamps go out. In order to meet the peak needs, nuclear power generation is maintained, however it is not easy to control the output at any time, so that too much electrical energy is sometimes produced. Accordingly, a pumped storage dam is a system often associated with nuclear power stations.

A new system shown in Fig.4.8 was proposed by the present authors in 1973[4]. Possessing no fossil fuels, which are a source of CO_2, solar and nuclear energies are chosen as the primary energies. One of the prominent features is that the primary energies are converted in multistage processes, that is, solar energy with shorter wavelengths is applied to generate electrical energy using solar cells and the unavailable (waste) energy is collected and applied to thermochemical water splitting sells to obtain hydrogen. Nuclear energy is utilized not only to get electrical energy from a nuclear reactor but also to obtain heat with a temperature of as high as 1,000°C which is then used to obtain hydrogen by applying it in a thermochemical water splitting cycle.

Although both Figs. 4.7 and 4.8 are similar, the first difference is, the existence of multistage systems. The next distinctive feature is that both the electrical system and the hydrogen system are closely connected by water electrolysis and the use of fuel cells. The present electrical power system uses a water pumping dam to store electrical energy, but the hydrogen energy system may reserve the electrical energy as hydrogen by electrolysis. A fuel cell is an effective facility for generating electrical power from hydrogen.

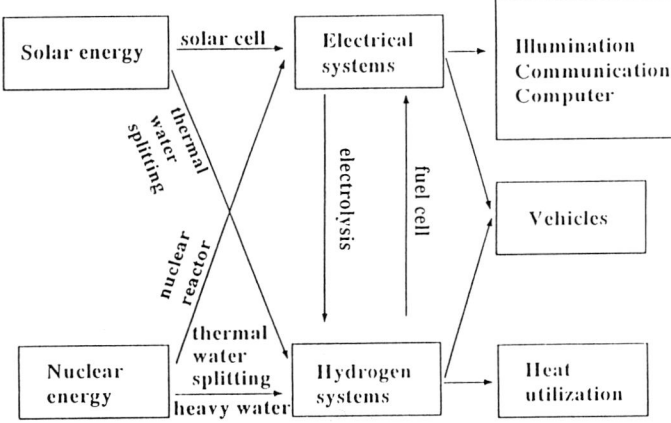

Fig. 4.8. Hydrogen energy system[4]

A utilization system can be designed so that the most suitable energy between electrical and hydrogen energy can be provided to the utilization subsystem.

Three prominent features of the **hydrogen energy system** are as follows: (1) It is a clean and non-depletable energy system. The term "clean" means there is no exhausted gas to contaminate the global environment, but if radioactive waste is regarded to play the same role as CO_2, then the hydrogen energy system should exclude the nuclear power source, so that only natural energy resources are used. (2) Hydrogen possesses a large amount of power, that is, its chemical wattage (heat evolved by the combustion per unit time) is the largest among the known chemical materials. Therefore hydrogen fuel has been chosen as a typical fuel for the second and third stages of rockets. (3) Hydrogen can replace petroleum when it is depleted. (4) Hydrogen is not only an excellent fuel but a basic chemical material to be synthesized into useful materials.

4-3. Energy Storage

(1) General survey : storage and transport

Energy storage as well as its transport plays an important role in energy systems in order to supply a stable amount of secondary energy, because primary energy resources are unevenly distributed and the need for the secondary energy is subject to heavy fluctuation.

Energy storage is most conveniently done by use of potential energy. Potential energy has two kinds of aspects, one of which is microscopic and the other macroscopic. The chemical cohesive energy of substances like fossil fuels and biomass fuels belong to the former. The chemical energy stored inside the fuels is the best way to store energy. The latter applies to dams, springs, electric or magnetostaticenergy, *etc*

The most difficult kinds of energies to store are photon and electromagnetic waves, the storage of which is usually carried out by conversion to other kinds of energy. Dynamic electric energy and heat energy are usually stored by conversion to mechanical potential energy (water pumping dam system) or chemical energy (water electrolysis). Storage using superconducting coils is under development.

The storage and transport of heat energy is basically difficult, because heat disperses by heat conduction, radiation and convection. However, excellent heat insulators, and materials that have very high reflection coefficients have been invented and apply to heat transportation for a distance as long as 10 [km]. Moreover, heat pipes have been developed so that heat is effectively transported. This will be described in a later chapter.

The ranking of the kinds of energies that are adaptable to storage is as follows: chemical energy, mechanical potential energy, electromagnetic potential energy, electric current, heat, electromagnetic wave, and photon energy.

The same ranking of adaptation to transport is decided by the maximization of

$$\Gamma = aML, \qquad (4.21)$$

where a, M and L are the quality factor, energy density, and transportation distance, respectivly. By the quality factor, the kind of transported energy is distinguished, and varies, case by case. It is preferable that it is the most conveniently adaptable.

Common fundamental conditions for storage and transport are as follows.

(i) *Prevention of any chemical reaction.* Fuels placed in contact with an atmospheric environment sometimes change in quality by oxidization or other reaction.

(ii) *Minimization of the energy loss due to dispersal processes (entropy production) generated by friction, discharge, radiation from a hot body, and other phenomena.*

(iii) *Prevention of the efflux of energy carrying materials (fossil fuels, heat medium, electric charge, etc.).*

All of these conditions are needed to minimize entropy production. This goal is difficult because the required technology is severely set against the second law of thermodynamics (entropy increasing law).

Representative practical facilities for energy storage are closely related to the strength of the constructing materials of storage vessels or instruments. The physics of this theory will be described later.

The main methods of energy transport are: (1) electric power line system, (2) pipeline system, and (3) container batch system, each of which will be discussed later.

To summarizes the effectiveness of energy storage and transport in practical systems are as follows:

(i) *Energy density.* Two kinds of energy density are introduced. They are volume density [kg/m^3] or [kcal/l] and weight density [kJ/kg] or [kcal/kg]. The former is more important for energy stations. The criterion of the effectiveness is taken as oil which has 9.8×10^6 [kJ/m^3]. No facility exists that has more capacity except for nuclear fuels, which have about 10^6 times the energy density of oil in uranium. However, the storage and conversion facilities are very complicated and expensive and are therefore not deemed superior to oil.

A hydrogen-air fuel cell has an output density a little smaller than oil. The energy density of a fly-wheel is also a little less than oil, but is not stable for long periods. We shall discuss these examples shortly.

In the case of vehicles (rocket, airplane, train, automobile), weight density is rather more important. Liquified hydrogen is the best fuel. Hydrogen gas of 1 [m^3] at standard state is first liquified. Then liquid hydrogen of 1.18 [l] and 83.76 [g] is obtained whose LHV is equivalent to gasoline with 0.32 [l] and 226 [g]. LPG and LNG are better than oil from the viewpoint of weight.

(ii) *Time factor (theory of pipeline)*. There is no explicit time factor in Eq.(4.21), but if it is rewritten as

$$\Gamma = aA\rho \int_0^T v \, dt \qquad (4.22)$$

$$\text{with} \qquad v = \frac{dL}{dt}$$

where A, ρ, and v are the cross-sectional area of the pipeline, the density of the energy carrier, and velocity of the carrier, respectively, and Γ is a function of T. The time T is the time necessary to reach the terminal at a distance L from the transport station. If a pressure P is supplied at the station, we have

$$\frac{dv}{dt} = \frac{P}{m} - \frac{v}{\tau}, \qquad (4.23)$$

where m is the mass density per unit length of pipe line, and $1/\tau$ is the coefficient of resistance due to viscosity and friction between the fluid and the inside wall of pipe. The necessary pressure to make a stationary flow is

$$P_\infty = (m/\tau) v_\infty \qquad (4.24)$$

However, the decay of stream is expressed by solving Eq.(4.23):

$$v = v_0 \exp[-t/\tau] + [1 - \exp(-t/\tau)](\tau/m) P. \qquad (4.25)$$

Time t appears explicitly in Eq.(4.25) but not in Eq.(4.24). If additional pressurizing stations are arranged to make a stationary flow, then we have no time factor. In the case of a batch system, if the successive transport is kept constant, then there is also no time factor.

(iii) *The time factor is rather important in energy storage.* Storage for too long a period yields unnecessary chemical reactions. The leaking of energy carriers is especially appreciable in the case of liquified fuels at very low temperatures. The temperatures of liquified hydrogen (which is sometimes denoted LH_2) and liquified methane (LNG) are 20.3 [K] and 111.8 [K], respectively. These are much too low to completely prevent vaporization. A device for recovering the vaporized LNG is equipped to LNG tankers and uses the recovered gas to drive electric power generators.

To minimize the vaporization of liquified fuels, the ratio of surface area to container volume should be minimized, that is, the vessel must be a sphere. On the other hand, spherical containers do not always utilize space efficiently. Some contradictions will always exist between economical and scientific requirements, but safety and environment should always be considered first.

(2) Practical systems

We shall survey energy storage systems or materials in which the storage densities are appreciable. Noticeable methods are shown in Table 4.2, where only those materials whose energy density is larger than oil are listed. On the other hand, storage systems have relatively small energy density and many of them are of small (*S*) and middle (*M*) scale except for pumping hydro power dam and compression gas systems (*L*).

Condensers, springs, spiral springs, and organic elastomers have too low an energy density to be used in actual macro-energy systems. Batteries, primary and secondary, are suitable for application to electrical and electronic equipments and necessities.

(a) Mechanical energy. There exist two kinds of mechanical energies that can be stored. One is kinetic and the other is potential energy.

(i) *Flywheel.* The kinetic energy of a body with its moment of inertia *I* rotating about its center of gravity with angular velocity *w* is expressed by

$$W = \frac{1}{2}Iw^2, \tag{4.26}$$

where we must notice that w is more important than *I*.

If the shape of the body is circular with radius *R*, and the total mass *M* is uniformly distributed at the rim, then

$$W = \frac{1}{2}MR^2w^2. \tag{4.27}$$

The stress σ acts along the circumference and is given by

$$\sigma = \rho R^2 w^2, \tag{4.28}$$

where ρ is the density of ring ($M = 2\pi R\rho$).
Thus we have, for the specific energy,

$$r = E/M = \sigma/(2\rho). \quad (4.29)$$

The coefficient 1/2 is determined by the body's shape and is replaced by K that can be larger than 1/2. In the case where K has a large value, the stress is applied along two directions and therefore the strength of the material is lowered. It is important to select materials with low density and high strength. Some typical materials are listed in Table 4.3[11,28].

Table 4.2. Density of stored energy by systems and materials (after A. W. Culp[11])

Materials [kJ/kg]		Systems [kJ/kg]		Scale
deuterium (D–D)	3.5×10^{11}	pumping up hydro power dam (100m)	9.8×10^{2}	L
uranium 235	7.0×10^{10}	silver oxide zinc battery	437	M
heavy water (fusion reaction)	3.5×10^{10}	nickel–hydrogen	160	M
natural uranium	5×10^{8}	lead–acid battery	119	M
95% Po – 210	2.5×10^{6}	flywheel	79	M
80% Pu –238	1.8×10^{6}	compressed gas	71	L
hydrogen (LHV)	1.2×10^{5}	organic elastomer	20	S
methane (LHV)	5×10^{4}	torsion spring	0.24	S
gasoline	4.4×10^{4}	coil	0.16	S
oil	2.8×10^{4}	condenser	0.016	S

Another definition of specific energy is given by

$$r° = E/V = K°, \qquad (4.30)$$

where V is the volume.

An engineering group in Switzerland manufactured a trolley bus which used a flywheel named "electrogyro" as early as 1950. Some trolley buses equipped with flywheels rotating at a maximum speed of 12,000 [rpm] were run in downtown San Francisco around 1976. The system of electrical power storage in the trolley bus is shown in Fig.4.9 [ref. *TIME*, **56** May 6, 1974]. Going down a downward slope, the bus drives the flywheel rotation up to 12,000 [rpm]. Going up an upward slope, the bus will get additional power from the flywheel generator.

Good materials such as alloys and epoxy glasses have energy densities as high as that of a lead-acid battery, but others are not competitive.

The flywheel system needs a vacuum space to avoid friction due to air viscosity and a frictionless support or suspension system. A levitation system utilizing the Meissner effect (a super conducting state which perfectly excludes magnetic flux) or strong magnetic forces has also been investigated.

Table 4.3. Flywheel materials (after A. W. Culp[11])

Materials	Density [kg·m^3]	Energy density [kJ/kg]	Working stress [10^6 Pa]	Cycle [1/s]
maraging-steel	8,000	130	7	10^4
4340 steel	7,833	64	4.3	10^4
Ti –6 –4	4,429	–	6.3	10^4
Al 2024 –T3	2,768	46	2.3	10^4
60 v/0 E –glass epoxy	1,993	128.1	6.2	10^5
63 v/0 Kevlar epoxy	1,356	–	10.0	10^6

136 Energy Technology

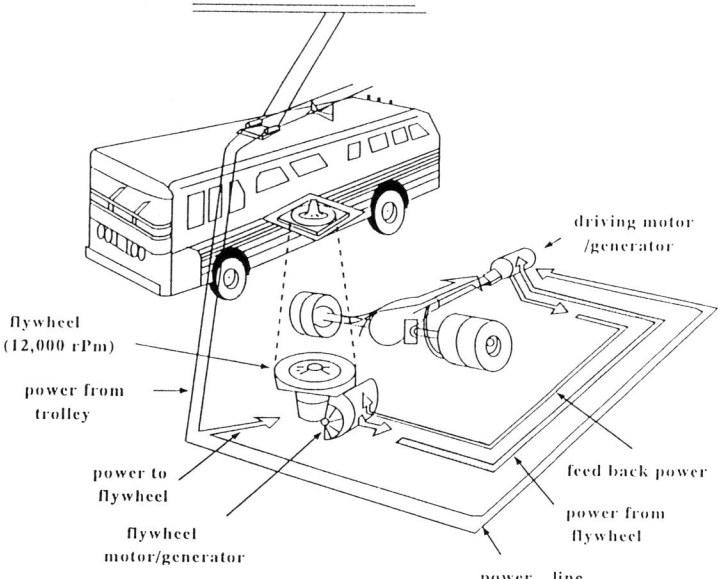

Fig. 4.9. Trolley bus equipped with flywheel.

A small scale storage of mechanical energy is the coil spring which provides the energy for driving a mechanical clock for several days if a low friction suspension is equipped.

(ii) *Pumping up hydropower dam*. This is such a long term that we call it sometimes "pumping up dam". It has been mentioned several times so far. This dam may be only the effective large scale electrical energy storage system at the present. Water of mass M [kg] is pumped up to a dam at the top of a hill h [m] high, then the stored energy is $E = Mgh = 9.8Mh$ [J]. It is also possible to hang a heavy body up at a high place instead of water, however such discontinuous processes may need many heavy bodies. In addition, maintaining security against earthquakes or an accident which cuts the supporting ropes is expensive and not competitive with the dam system.

Recently, the location of pumping up dams at coastal sites has been examined, and utilizes sea water. The site selection is easier but countermeasures against erosion caused by sea water are necessary.

(iii) *Gas-compression*. A power storage system combining an airtight underground cavity with a pond on the ground is shown in Fig.4.10. Such systems are seen in Switzerland, Sweden, and U.S.A. We shall explain a Swedish system (Sydsvenska Kraft)[1,15] using the figure. First, the compressor (1) is driven by a motor (2) to compress the air up to 15×10^3 [hPa] and cooled by the cooler (6) and compressed into the underground cavity with a volume of 28×10^4 [m^3]. Water

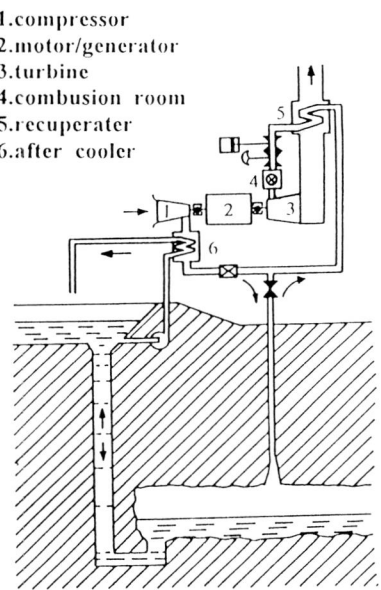

1. compressor
2. motor/generator
3. turbine
4. combusion room
5. recuperater
6. after cooler

Fig. 4.10. Structure of gas compression power storage[15].

filled in the cavity is pumped up to the pond on the ground that is higher than 150 [m] from the cavity. If electrical power is needed, then the valve is opened so that the compressed air comes up to a recuperator (5) where the air is modulated to adequate pressure and is forwarded to a combustion room (4). The air from the combustion room has high temperature as well as high pressure which is enough to drive the gas turbine thus generating electrical power. This example system is equipped with a compressor of 70×10^3 [kW] and the output of electric power is 700×10^3 [kW].

The reason why such gas-compression power storage is possible is that the vessel which endures high pressure is replaced by a cavity. If a high pressure vessel is applied, the density of stored energy will be very small, because the weight of the vessel is very heavy and the volume cannot be large enough to overcome the weight limitation.

Note that the stored energy differs according to the choice of gas. Work needed to make adiabatic compression of 1 [mol] gas is given by, referring to Eqs. (2.2) and (2.6),

$$W = R/(T_1 - T_2)/(\gamma - 1) \quad (<0)$$

$$= C_v T_1 [1 - (P_2/P_1)(\gamma - 1)/\gamma] \qquad (4.31)$$

where suscripts 1 and 2 refer to before and after the compression. Equation (4.31) shows that gases with large γ and C_v are profitable and the ratio P_2/P_1 needs to be as large as possible.

Assuming that the efficiencies of the gas compressor and the generator are η_c and η_g, respectively, and the necessary work to store the energy is W/η_c, then the overall output is given by $(Q + W/\eta_c)\,\eta_g$, where Q is the heat input.

(iv) *Metal hydride.* Some alloys can absorb hydrogen gas at more than 700 times the alloy's volume. This reversible chemical cycle is

$$2M + H_2 \rightleftarrows 2MH + Q, \qquad (4.32)$$

where M is the alloy and Q is the reaction heat with order of magnitude 10 [kcal/mol-H_2]. The pressure of the hydrogen gas and temperature of the alloy are sensitively related to the component ratio of the hydrogen number density to alloy atoms. The characteristic behavior is called *PCT*-curves. This reversible reaction is utilized not only to store mechanical energy but also to store heat as well as hydrogen. One can readily understand that the compressed hydrogen is forced to be stored in the alloy generating the evolved heat.

When hydrogen is absorbed by the alloy then this can be regarded as storing mechanical energy, because if the alloy is warmed up, high pressure hydrogen gas is recovered. This phenomenon is utilized as a compressor.

The metal and metal hydride cycle is very important, as has been mentioned in Table 2.1, and will be introduced, in detail, in a later section.

(v) *Cryogenic storage.* Normal butane (0.6°C), propane (- 45°C), ethane (- 90°C), and methane (-161°C) are excellent fuels but are gaseous under atmospheric pressure. The liquefied gas has a much higher energy density and the said gases are distributed in the liquid state as LPG (liquefied petroleum gas) and LNG (liquefied natural gas).

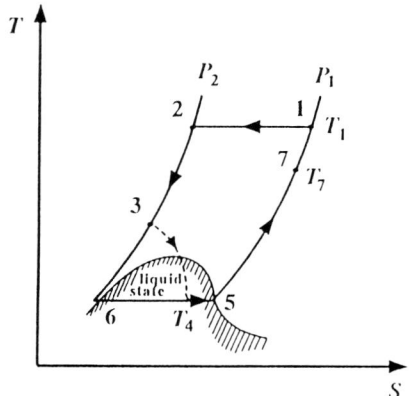

Fig. 4.11. Temperature *vs* entropy curves in liquefaction cycle.

Liquefaction of gas, which needs mechanical energy and cooling heat energy, is a kind of storage of mechanical energy. Temperature *vs* entropy curves in the liquefaction cycle are shown in Fig. 4.11. Starting from pressure P_1, atmospheric pressure, the medium gas is isothermally compressed to the higher pressure P_2 ($> P_1$). Then the gas is cooled down to a state 3 where the entropy is small and the gas in state 3 is adiabatically expanded to state 4, the liquefied state, which is shown with shading. A part of the liquid is vaporized, and the temperature is lowered even more. The cryogenic liquid is used to cool the gas medium to state 7 (pressure P_1, temperature $T_7 < T_1$) and a similar process is undertaken, making it easier to obtain the liquefied state[50].

The usual air-liquefier invented by C. von Linde (1843 - 1934) has the following processes. (1) Air is compressed from atmospheric pressure to 65 x 10^3 [hPa] being cooled by water to 0°C. (2) Air at this state (65 x 10^3 [hPa] and 0°C) is expanded to 22 x 10^3 [hPa] and -11°C. (3) Isothermal compression and adiabatic expansion are repeated to attain -140.7°C where air is liquefied at 22 x 10^3 [hPa]. (4) The liquefied gas is expanded to atmospheric pressure to attain -196°C. (5) This liquid is stable under the atmospheric pressure. Strictly speaking, the atmospheric boiling temperatures of oxygen and nitrogen are -183.0°C and -195.0°C, respectively.

Hydrogen and helium liquefaction were first achieved by J. Dewar (1842 - 1923) in 1898 and by H. Kamerling-Onnes (1853 -1926) in 1908, respectively.

The rocket engine RL 10, which utilizes LH_2 as propellant, was developed over seven years, from 1956. The quantity of LH_2 consumed during that period amounted to one million liters[50].

Table 4.4. Enthalpy, work and cooling heat energy for gas liquefaction

Gas	Boiling point [K]	Cooling heat [kcal/kg]	Work [kJ/kg]	Total energy [kJ/kg]
ethylene	169.4	155.5	470	1118
methane	111.8	218.0	1147	2025
oxygen	90.2	98.4	631	1041
nitrogen	77.3	91.5	777	1200
hydrogen	20.4	923.3	12190	16037
helium	4.2	367.2	8389	9919

We have learned that gas liquefaction needs not only mechanical energy but also cooling, therefore the total energy is given by

$$H = W + (-Q), \qquad (4.33)$$

where $-Q$ is the cooling heat quantity (positive exergy).

Theoretical values of H and $-Q$ for some gases are shown in Table 4.4. Note that hydrogen needs more than 10 times the cooling heat of nitrogen because its specific heat is very large. When liquefied hydrogen is heated to the gaseous state, its expansion rate is up to 865 times, and it is therefore often applied to rockets, jets and engines of surface vehicles.

Lastly, we shall note that solid hydrogen is produced at 13.8 [K] under atmospheric pressure. A slurry of solid and liquid hydrogen is considered to be a fuel for rocket and jet engines, because its lifetime at the low temperature state is longer than that of the liquid state.

Metallic hydrogen is theoretically realized at a critical pressure of 1.6×10^9 [hPa]. A Russian group[34] has observed metallic hydrogen at the super high pressure of 2.8×10^9 [hPa]. Their experimental procedure is as follows. Solid hydrogen is placed under a super high pressure that is gradually increased until a sudden change of density from 1.08 [g/cm^3] to 1.6 [g/cm^3] is observed when 2.8×10^9 [hPa] is applied.

The volumetric energy density of metallic hydrogen is as large as 18.3 times that of LH$_2$ and 46 times that of gasoline. There exists no comparative fuel other than nuclear fuels.

(b) Electrical energy. Electrical energy is stored as chemical energy in batteries. Batteries are classified into two groups, one of which is the primary battery and another is the secondary battery. The former is not rechargeable, while the latter is rechargeable. The fuel cell is sometimes included as a secondary battery, however, it is actually a kind of generator having no storage function of electrical energy.

(i) *Conversion from chemical energy to electrical energy.* The Gibbs' free energy G, which is defined by Eq.(1.41), of a material is given by

$$G = U + pV - TS. \qquad (4.34)$$

If G is subject to a change under isothermal and isobaric conditions, then we have, from Eq.(4.34),

$$(\Delta G)_{p,T} = \Delta U + p\Delta V - T\Delta S, \qquad (4.35)$$

where the change of internal energy U is the balance between the input heat and the output work, *i.e.*,

$$\Delta U = Q - W \qquad (4.36)$$

Substituting Eq.(4.36) into Eq.(4.35), we get

$$(\Delta G)_{p,T} = Q - W^\circ - T\Delta S \qquad (4.37)$$

with
$$W^\circ = W - p\Delta T. \qquad (4.38)$$

The quantity W° is the net work, from which the mechanical work done against the outside atmosphere is subtracted and called **"effective work"**. It is this W° that can be taken out as electrical energy from the chemical material.

On the other hand, we have from the second law of thermodynamics,

$$S > Q/T \quad \text{(for irreversible)} \qquad (4.39)$$

and
$$S = Q/T \quad \text{(for reversible process)} \qquad (4.40)$$

Therefore we have

$$(\Delta G)_{T,p} + W' \leq 0 \qquad (4.41)$$

i.e.,
$$W' \leq -(\Delta G)_{T,p} \qquad (4.42)$$

Equation (4.42) shows that the upper limit of energy that can be taken out of material is equal to its Gibbs' free energy. Conversely the increment of Gibbs' free energy DG is the minimum electrical energy required by electrolysis.

The maximum efficiency of the conversion from chemical to electrical energy is given by

$$\eta_m = (\Delta G/\Delta H) = 1 - T(\Delta S/\Delta H) \qquad (4.43)$$

Let the maximum voltage be V°, then

$$V^\circ = \Delta G/nF, \qquad (4.44)$$

where n and F are the number of transferred electrons and Faraday's constant, respectively.

For example, one [mol] of hydrogen reacts with 1/2 [mol] of oxygen to generate one [mol] of water, then $\Delta G = 56.3$ [kcal/mol] at 298 [K] and 10^3 [hPa] and $n = 2$, so that $V^\circ = 1.22$ [V].

The total amount of energy that can be taken out of the battery is

$$W = V^\circ I t, \quad [J]$$

where I [A] and t [s] is the extracted electrical current and the sustainable time, respectively.

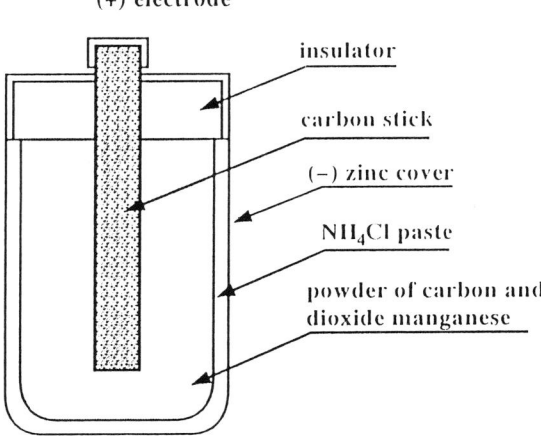

Fig. 4.12. Structure of manganese dry battery

(ii) *Primary battery*. This type of battery has been distributed widely all over the world since C. Gassner created the cylindrical "dry cell". He applied G. Leclanche's idea in 1868 to create a usable and convenient structure. The world market of dry cell batteries (manganese 65 %, alkali manganese 22 %, silver dioxide 11 %, and others 11 %, in 1985) was 3.78 billion dollars, excluding Russia and China where statistics were unavailable[48].

An outline of the structure of a manganese dry cell battery is shown in Fig. 4.12. The chemical reaction that is caused by connecting both electrodes is

$$MnO_2 + 2NH_4Cl + Zn \rightarrow$$
$$MnO + Zn(NH_3)_2Cl_2 + H_2O, \qquad (4.45)$$

where two electrons transfer along the path as follows :

$$\begin{array}{l} Mn^{4+} + 2e^- \rightarrow Mn^{2+} \text{ (anode)} \\ \downarrow \\ \text{circuit} \\ \downarrow \\ 2e^- + Zn^{2+} \leftarrow Zn \text{ (cathode).} \end{array} \qquad (4.46)$$

The open circuit voltage is about 1.5 [V] in the case of Eq.(4.46).

The primary battery is one of the more sophisticated products that human beings invented at an early age but to which no radical improvement has yet been made.

(iii) *Secondary battery.* The most popular secondary battery is the lead-acid battery that is widely utilized to help start internal combustion engines. The electrochemical mechanism is

$$\text{Pb(-Sb)} \mid \text{Pb} \underset{\text{charge}}{\overset{\text{discharge}}{\rightleftarrows}} \text{PbSO}_4 \text{ (aq)} \underset{\text{charge}}{\overset{\text{discharge}}{\rightleftarrows}} \text{PbO}_2 \mid \text{Pb(-Sb)} \qquad (4.47)$$

The above mechanism is explained with the aid of Fig.4.13 (a). In the case of discharge, Pb becomes ionized at the cathode, *i.e.*,

$$\text{Pb} \rightarrow \text{Pb}^{++} + 2e^-,$$

two electrons go to the anode *via* the external circuit, and Pb^{++} reacts on the sulfuric radical SO_4^{--} to produce PbSO_4. Two hydrogen ions 2H^+ from sulfuric acid react on PbO_2 as

$$2\text{H}^+ + \text{PbO}_2 \rightarrow \text{Pb} + \text{H}_2\text{O} + \frac{1}{2}\text{O}_2$$

at the anode.

In the charging process, electrons are removed so that the reaction :

$$\text{Pb} + 2\text{H}_2\text{O} \rightarrow \text{PbO}_2 + 4\text{H}^+$$

occurs, and the generated 4H^+ proceed toward cathode and the reaction

$$2\text{PbSO}_4 + 4\text{H}^+ \rightarrow 2\text{Pb} + 2\text{H}_2\text{SO}_4$$

takes place.

As another example, a Ni-H_2 battery is considered. This type is rapidly spreading in place of Ni-Cd batteries, because Cd is a source of public pollution. The cathode is a metalhydride like MmNi_5H_6 (Mm denotes mischmetal which is a mixture of alkali earth elements). The electrochemical mechanism is shown in Fig.4.13 (a), (b), and (c). First, we shall explain the discharging process. A hydrogen ion H^+ is emitted to the KOH electrolyte from the cathode and the electron goes to anode *via* the external circuit. At the anode, the attained electrons is caught by NiOOH *via* the reaction :

$$e^- + \text{H}^+ + \text{NiOOH} \rightarrow \text{Ni(OH)}_2.$$

If the NiOOH is depleted, then a charging process is necessary. (c). An electron is removed from the Ni(OH)_2 of the anode, and a reverse reaction occurs to yield NiOOH and H^+ which goes into the cathode, reacting to produce metal hydride.

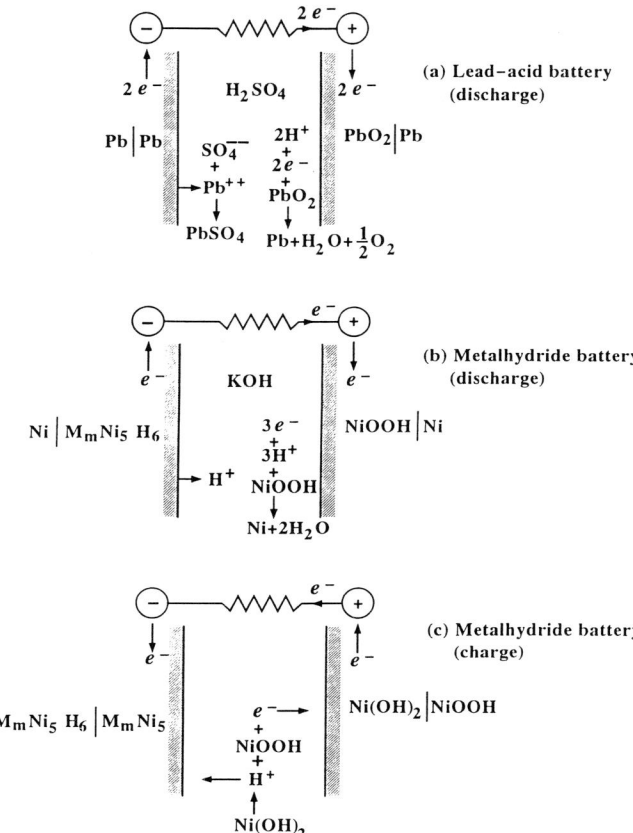

Fig. 4.13. Electrochemical mechanism of a secondary battery.

The merits of such Ni-H$_2$ batteries are: (1) the mobility of H$^+$ is larger in the KOH electrolyte than in H$_2$SO$_4$, (2) the reaction at the cathode is simple, (3) metal hydrides can absorb large amounts of hydrogen that can be stored stably for long periods. The only demerit is considered to be the price because mischmetal is expensive.

The capacity of typical secondary batteries is shown in Table 4.5.

A mischmetal nickel hydride battery that was manufactured recently in the U.S.A. is compared to the U.S. Advanced Battery Consortium in Table 4.6 [S.R. Ovshinsky *et al.*, *Science* **260** (1993) 176]. According to these data, this new type of battery is superior to Ni-Cd, and other batteries.

(iv) *Fuel cell*. A fuel cell is a system in which the stored chemical energy of fuel is converted directly to electrical energy. "Directly" means not *via* heat energy. Fuel cells do not go through the thermal regime and the conversion from fuel to electrical energy is not subject to the limit of Carnot's efficiency. The electrons of fuel transfer directly to oxygen *via* an external circuit. No exhausted gas is emitted production is also very small. The amount of available exergy is much more. For these reasons, considerable interest has been generated and much effort has been invested into the development of this system.

The hydrogen-oxygen fuel cell is a typical and practical example. The cell is composed of cathode, anode, and electrolyte (usually, potassium hydroxide, KOH, or phosphoric acid H_3PO_4). The electrolyte is closed between the both electrodes. Hydrogen and oxygen is fed to the cathode and anode, respectively. The mechanism of generating electric current is shown in Fig.4.14. The Pt (platinum, catalyzer is attached to the cathode where two electrons ($2e^-$) are removed from the outermost electronic shell (6S) of the Pt atom which receives two electrons from hydrogen, H_2. The catalytic function is the mediation of electron-transfer between the electrode and fuel.

Table 4.5. Capacity of secondary batteries

	Type	Capacity		Life time [cycle]
		energy density [Wh/kg]	energy density [Wh/l]	
traditional	lead–acid	40 –50	8 –150	1000 –1500
	Ni –Cd	25 –150	50 –80	1500 –3000
	Ag –Zn	100 –150	200 –280	50 –150
inovated	Zn –air	80 –120	100 –200	–100
	Na –S	330	490	–
	Li – Cl	300 –500	500 –800	–
	Li –halogen –ide	200 –500	100 –250	–
	Li –polymer	90 –110	350	–

Wh = 3.600 kJ = 0.86 kcal

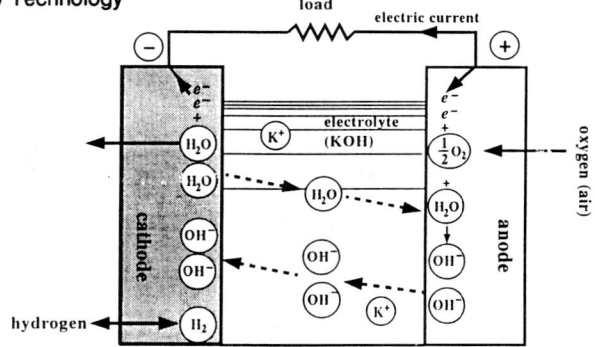

Fig. 4.14. Function diagram of a fuel cell. The Pt catalyzer emits electron e^- to the cathode and receives an electron from the fuel, H_2. Electrons and air (oxygen) react on water generating OH^- ions.

The chemical reaction at the cathode is

$$(Pt) + H_2 \rightarrow 2H^+ + 2e^- \quad \text{(to external circuit)} \tag{4.48}$$

and
$$2OH^- + 2H^+ \rightarrow 2H_2O$$

where two hydroxyls are given from the anode. The two electrons do electrical work at the load on way to the anode.

The chemical reaction at the anode is

$$2e^- + \frac{1}{2}O_2 + H_2O \rightarrow 2OH^-, \tag{4.49}$$

where the generated hydroxyls go to the cathode. Comparing Eqs.(4.48) with (4.49), we know that one water molecule is generated in excess during the reaction. Excess water is used for drinking and cooking in the case of the Apollo spaceship.

Hydrogen fuel is not burned in the fuel cell so that the amount of available exergy is large and entropy production is very small. According to Eq.(4.43), the efficiency of the fuel cell is much higher when compared to internal or external combustion electric generators. The efficiency of fuel cells reaches more than 40 %, which is the highest among other engines and only gas turbines working at temperatures higher than 1,300°C are competitive. The comparison is shown in Fig.4.15.

If it is possible to find a good catalyzer, any fuel can be fed to the cathode of the fuel cell to generate electrical power. The open circuit voltage differs according to the fuel. Some examples are as follows: hydrogen - 1.23 [V], methane - 1.2 [V], propane - 1.09 [V], carbon monoxide - 1.33 [V], and gasoline - 1.21[V].

Table 4.6. Comparison of battery performance of a realized metal hydride battery to USABC (U. S. Advanced Battery consortium) (after S. R. Ovshinsky *et al.*) *Science*, **260** (1993).

Property	USABC	Metalhydride Battery
Specific energy (Wh kg^{-1})	80(100 desired)	80*
Energy density (Wh per liter)	135	215*
Power density (W per liter)	250	470
Specific power (W kg^{-1}) (80% DOD in 30s)	150(>200 desired)	175
Cycle life (cycles)(80%DOD)	600	1000
Life(years)	5	10
Environmental operating temperature	−30 to 65°C	−30 to 60°C
Recharging time	<6hour	15min(60%) <1 hour (100%)
Self discharge	<15% in 48 hours	<10%in 48 hours

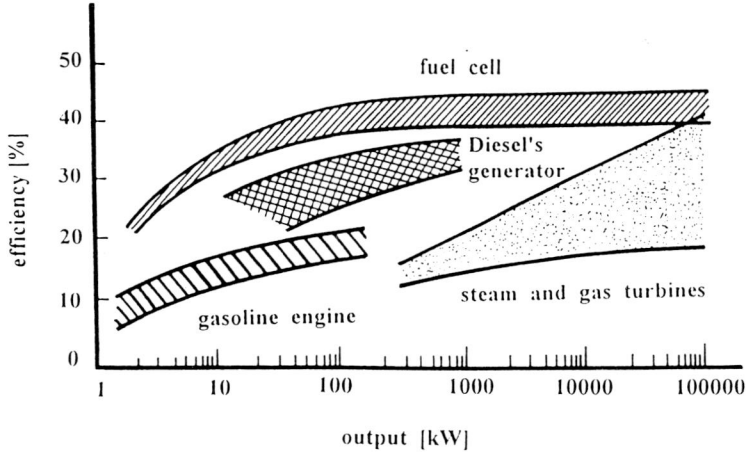

Fig. 4.15. Comparison of electrical power generators. Gas turbine temperatures of as high as 1,300°C can compete with a fuel cell.

Table 4.7. Classification of fuel cells

	PAFC	SOFC	MCFC
electrolyte	$H_3PO_4 + H_2O$	Zr-ceramics	Litium carbonate
carrier	H^+	O^{2-}	CO_3^{2-}
temperature	200°C	1,000°C	650°C
fuel	H_2	H_2, CO	H_2, CO, coal gas
efficiency	40%	50%	50%
comments	reliable	· internal reforming of fuel · heat-resisting ceramics	· internal reforming of fuel · erosion
diagram of generating plinciples		(SOFC diagram: H_2, CO and H_2O, CO_2 at fuel side; O^{2-} electrolyte; $\frac{1}{2}O_2$ at air side; load)	(MCFC diagram: H_2, CO and H_2O, CO_2 at fuel side; CO_3^{2-} electrolyte; $\frac{1}{2}O_2$, CO_2 at air side; load)

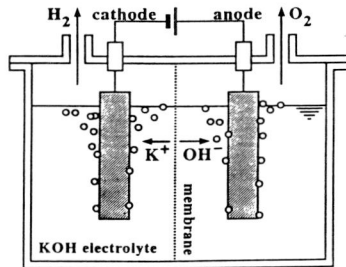

Fig. 4.16. Mechanism of water-electrolysis. The potassium ion, K^+, of the electrolyte is attracted to the cathode, where one proton H^+ of the water molecule (H_2O) is removed and the remaining OH^- goes towards the anode.

Fuel cells that have been developed are classified into three types. The first, the hydrogen-oxygen (air) type fuel cell introduced above, whose electrolyte is phosphoric acid (**PAFC**), has been developed in recent years. The Tokyo Electric Power Co. has developed such a fuel cell with a 11×10^3 [kW] capacity that works at temperatures ranging from 50 - 220°C.

The next generation of fuel cells work without noble catalyzers such as Pt or Ag-Ni alloy. An example of this type is the molten carbonate fuel cell (**MCFC**) whose fuel is carbon monoxide or hydrogen. It works at high temperatures of 650 - 700°C.

The third generation of fuel cell is the solid electrolyte fuel cell (**SOFC**). Ceramics of zirconium are applied as the electrolyte and the fuel is hydrogen or carbon monoxide. It works at temperatures as high as 1,000°C.

A comparison of the three types of fuel cells is shown in Table 4.7, where the basic mechanism of generating electricity via the MCFC and SOFC is given. The reactants of these fuel cells are H_2 and CO produced by reforming CH_4, methanol, and coal gas. The reformer is found outside the cell in PAFC, while in the MCFC and SOFC it is possible to reform the fuels to produce the reactants in the interior of cell.

Among the three kinds of fuel cells, PAFC is the most practical at present and waste heat from it has a temperature greater than 100°C. Accordingly development of co-generation (ref. p.126) with use of this waste heat is flourishing. If both conversions are taken into account, the system efficiency can attain 70 %.

Fuel cells are, undoubtedly, at the frontiers of energy conversion.

(v) *Water electrolysis*[37]. An important elemental technology of the hydrogen energy system (Fig.4.8) is water electrolysis, which is a partner of fuel cells and traditional technology. The principle of water electrolysis is quite the reverse of the fuel cell, as shown in Fig.4.16.

150   Energy Technology

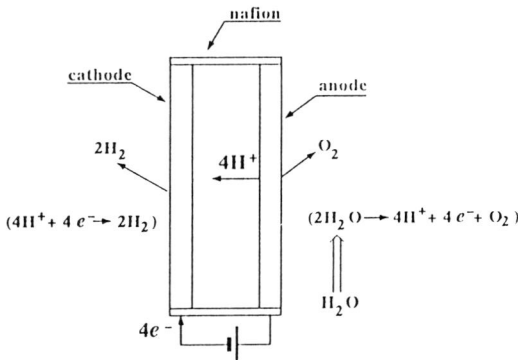

Fig. 4.17. Principle diagram of SPE (solid polymer electrolysis). Nafion is the name of GE's merchandise.

Water in the electrolyte (*e.g.*, KOH, NaOH, H_2SO_4) is ionized into $2H^+$ and $2OH^-$ and

$$2H^+ + 2e^- \rightarrow H_2 \quad \text{(at cathode)},$$

where the two electrons come from the anode where the reaction:

$$2OH^- \rightarrow \frac{1}{2}O_2 + H_2O + 2e^- \quad \text{(at anode)}$$

takes place.

The necessary voltage for electrolysis is theoretically the same as the generated voltage of the hydrogen-oxygen fuel cell, which is 1.22 [V]. However, collisions with other ions, molecules, and bubbles cause reistance for the OH^- ions to move through the electrolyte. The bubbles near the electrode surfaces have relatively large resistance. Such resistance yields excess voltage to electrolyze water, the so-called "over-voltage". In the case of conventional water electrolysis, about 2 [V] are needed to carry out electrolysis. The over-voltage is 0.77 [V] (= 2 - 1.23 [V]).

The necessary Gibbs' free energy (electrical energy) is given by

$$\Delta G = \Delta H - T\Delta S, \tag{4.50}$$

where ΔH and ΔS are the changes of enthalpy (total energy) and entropy, respectively, and a reversible isothermal change is assumed. Because the necessary total energy ΔH is composed of Gibbs' free energy ΔG and heat energy $T\Delta S$, less Gibbs' free energy is required as the temperature increases. The free energy in this case is electrical energy which is more valuable than heat energy and methods are investigated for reducing ΔG by increasing T. For this purpose, a higher temperature electrolysis and water vapor electrolysis have been developed.

Electrolysis at higher temperature has been eagerly studied. A typical electrolyzer is the solid polymer electrolyzer (SPE), which electrolyzes water at temperatures as high as 150°C and 2×10^5 [hPa]. Very high performance is achieved if Nafion is applied as the electrolyzer. Figure 4.17 shows the principle diagram of SPE. Such a SPE has small scale, *i.e.*, a unit cell that is designed to apply 2 [A/cm^2] has only 0.25 [mm] thickness.

Besides SPE, VPE vapor phase electrolysis using Zr-It or LaMnO$_3$-Sr alloy have been investigated as mentioned before.

Another important point is the catalyst. The catalytic factors involved in the H$_2$ evolution reaction have been classically explained in terms of heats of absorption of hydrogen and percentage d-character of the catalytic metal (in other words, these are the mediation of valence electrons). Almost all the catalysts for the H$_2$-evolution electrode have been high surface area nickel based alloys such as intermetallic compounds of Ni, Ni-Mo, Ni-Co, Ni-Fe, Ni-Mo-Cd, Raney nickel, and *etc*. The pathways involved in the oxygen evolution reaction are more complex than those proposed for H$_2$-evolution since the discharge of O$_2$ occurs on oxide covered or non-metallic surfaces. In recent years, a number of mixed transition metal oxide catalysts have been developed. Of these, NiCo$_2$O$_4$ and Li doped Co$_3$O$_4$ appear promising.

The minimum energy consumption rate of H$_2$-production by electrolysis is about 4.5 [kW·h/m^3-H$_2$] at the present stage.

We shall list some typical industrial water electrolysis-facilities worldwide:
1. Asahi Glass (Japan): activated cathode/ Re-Rh modified Raney nickel, 1.7 [V], 70 [A/dm^2], 110°C, 30 wt % KOH. 2. Jülich KFA Electrolyzer (Germany); Raney nickel activated by Lye treatment, 1.5 [V] at 4 [kA/m^2], 100°C, 10 [M] KOH. 3. Life Systems, Inc. (U.S.A.); 1.5 [V], 1.24×10^4 [hPa], 0.16 [A/cm^2]. 4. NORSK Hydro (the Netherlands); 1.8.[V], 2.5 [kA/M^2] (4.3 [kEh.m^3]), 80°C, 25 % KOH (4.4 [kW·h/m^3]).

(vi) *Condenser and coil.* Condenser and coil storage are not available for practical power utility usage yet but are applied very often in electrical and electronic circuits. These stored energies are expressed by Eqs.(1.14) and (1.18), respectively. If a large current is instantly needed, then current from a condenser is useful. A large scale storage system of electrical energy using superconducting wire has been investigated. The stored energy density of this system is considerable but the tension of the structural facility supporting the superconductors must be strong enough, and this will be discussed in a later section.

(c) **Heat energy.**

(i) *General discussion.* Heat energy is determined by the temperature difference between the heat medium and the environment. Higher temperatures as well as lower temperatures are the object of heat energy storage. The latter has already been mentioned, so that storage of heat at high temperature is discussed here.

152   Energy Technology

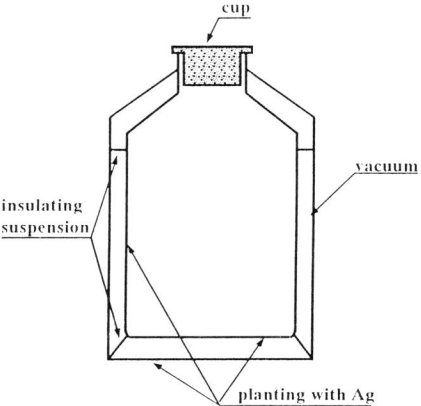

Fig. 4.18. Thermos flask.

Heat with higher temperature tends toward equilibrium with the environment by the processes (i) radiation, expressed by Eq.(1.33), (ii) convection, which is realized in fluids, (iii) heat conduction that occurs in any material except for vacuum.

A typical heat storage vessel is the "thermos flask" (vacuum bottle) structure shown in Fig. 4.18. Heat insulation is devised by providing the double structure, *i.e.*, the inner bottle is suspended by heat insulating thin supports and the space between the inner bottle and the inner wall of the outer bottle is evacuated. In addition, both surfaces of the inner bottle and the inner surface of the outer bottle are plated with silver giving a high reflective coefficient. As a result of such a design, escape of heat is prevented in three ways to stop the heat dispersion described in (i), (ii), and (iii) above. A thermos flask represents the principle of heat storage.

However, strictly speaking, the term "heat" used above means sensible heat. Heat is most conveniently stored by using latent heat.

Examples of data for sensible- and latent-heat storage are shown in Table 4.8, and 4.9, respectively.

Table 4.8. Sensible heat storage

Material	Chemical formula	Energy density [kJ/kg°C]
water	H_2O	4.18
ethyl alcohol	C_2H_5OH	2.85
limestone	$CaCO_3$	0.91
sand	SiO_2	0.80

Table 4.9. Latent heat storage

Material	Chemical formula	Energy density [kJ/kg°C]	Melting point[°C]
water	H_2O	335	0
sodium sulfate decahydrate	$Na_2SO_4 \cdot 10H_2O$	153	32
calcium hydroxide tetrahydrate	$Ca(NO_3)_2 \cdot 4H_2O$	153	42
barium hydroxide octahydrate	$Ba(OH)_2 \cdot 8H_2O$	300	78
sodium	Na	116	98
ammonium thiocyanate	NH_4CNS	261	146
sodium hydroxide	NaOH	226	300
lithium hydride	LiH	3,780	685
sodium chloride	NaCl	493	810
sodium sulfate	Na_2SO_4	163	884
glycerol	$C_3H_8O_3$	201	18

The general expression for heat energy storage by a heat medium can be written as

$$Q = m[\int_{T_1}^{T_m} C_s dT + \int_{T_m}^{T_b} C_l dT + \int_{T_b}^{T_2} C_g dT$$

$$+ \Delta H_{sl} + \Delta H_{lg} + \Delta H_c] \tag{4.51}$$

where Q is the heat energy that is stored by m [mol] medium, T_l, T_m, T_b, and T_2 are the lowest, melting, boiling, and highest temperatures, respectively, C_s, C_l, and C_g are the heat capacities at constant pressure of solid, liquid, and gas, respectively. Latent heats ΔH_{sl}, and ΔH_{lg} are those between solid and liquid, and between liquid and gas, respectively. Heat storage by chemical reaction is introduced in Eq.(4.51) by the term ΔH_c that will be discussed below.

From Tables 4.8 and 4.9[11,22], it is obvious that water is the best sensible heat storage medium. As for high temperature heat storage, most heat media other than water are corrosive, and hence water is the best medium for heat storage. This is one of the main reasons that the earth's atmosphere is mildly activated by the heat cycle due to water.

Note that storage at high temperature is difficult. Radiation from a body with temperature T is given by Eq.(1.33) so that if the temperature rises by ΔT then the increase of radiation energy is

$$\Delta W = 4\sigma T^3 \Delta T. \tag{4.52}$$

The radiation from a body at 500°C is about 8.9 times that at 100°C. Preventing radiation from a body at high temperature is very difficult.

(ii) *Storage by chemical reaction.* If the following types of reversible thermochemical reaction are possible, then we can store heat stably by chemical reactions. The necessary reversible chemical reactions are:

$$1.\ A + \text{heat} \rightleftarrows B + C, \tag{4.53}$$

$$2.\ P + Q + \text{heat} \rightleftarrows R + S \tag{4.54}$$

Equation (4.53) indicates the thermal dissociation of substance A into B and C. If heat is needed, the chemical reactants B and C are reacted with each other to produce heat. Examples of this kind of heat storage are

1-1. dilute sulfuric acid + heat
$\rightleftarrows$ concentrated sulfuric acid + water (4.55)

1-2. $CaCl_2 \cdot 6NH_3$ + heat
$\rightleftarrows$ $CaCl_2 \cdot NH_3 + 5NH_3$ (4.56)

1-3. $NiCl_2 \cdot 6NH_3$ + heat
$\rightleftarrows$ $-NiCl_2 \cdot 2NH_3 + 4NH_3$ (4.57)

When concentrated sulfuric acid of 65 wt.% is dissolved into water and mixed, the temperature is raised by 20°C. The heat storage processes expressed by the reaction from left to right in Eqs.(4.56) and (4.57) indicate the dissociation of ammonia molecules from ammonium compounds by heat. The heat evolving processes are shown by the reverse arrow. Nickel dichloride of 1 [kg] stores heat of 600 [kJ/kg] at 175°C under atmospheric pressure. However, the material is so corrosive at high temperature and the heat conductivity less than good that efficient heat storage and evolution are not possible in practice.

Another example is

1-4. $Ca(OH)_2$ + heat $\rightleftarrows$ $CaO + H_2O$ (4.58)

which is carried out at 500°C.
A well-known example of Eq.(4.54) is

$$CH_4 + H_2O + \text{heat} \rightleftarrows CO + 3H_2,$$ (4.59)

This reaction is called "EVA-ADAMS" and has been a German project. This is the process of steam-reforming of methane and advances from left to right at temperatures as high as 850 - 1200°C and the heat evolving reaction occurs at 350 - 700°C. The heat storage density is 120 [kJ/(kg mol CH_4)].

However, both reactants of Eq.(4.59) are in a gaseous state so it is difficult to separate them from each other. Therefore the heat storage is carried under mixed gas conditions.

Heat storage and a heat pump using the metal hydride cycle will be introduced in detail later in this chapter.

(d) Principle of energy storage.

Energy storage is an essentially important factor in energy systems and the examples of the pumping up dam, flywheel, elastic body, gas-compression, oil, batteries and others have been discussed so far. Storage is carried out by either materials, such as the case of heat storage, or structures, as a pumping dam, flywheel, *etc.* States of energy storage are in a non-equilibrium state, that is to say, an energy storage system is a sustainable structure of materials or structures in an unstable state. Chemical energy is exceptionally adaptable to storage because chemical substances have cohesive energies that are confined stably within them. These are obtained out by a relatively small ignition energy.

The other energies enumerated above have a common aspect in that their storage mechanism is due to a sustained structure, which is represented by a storage vessel. H. Takahashi[46] has proposed general rules that govern the structure or vessel of energy storage.

(i) *The total amount of storable energy is determined by the mechanical strength of the material that composes the structure of the storage system.*

For examples, the pumping up hydro dam has a capacity of storable water maintained by a dam bank. The quantity of stored water is determined by $M = A \times h$, where A and h are the dam area and the height. The effectiveness is expressed by the height of dam relative to the power generation station. It is obvious that a higher dam needs a stronger bank. It is a matter of course that the bank shape is also designed to enhance the ultimate strength of the bank.

Energy storage by gas-compression needs a natural cavity or artificial vessel. The limit of the storable compressed gas in both systems is also determined by the stress of the wall of the storage structure.

In the case of artificial vessels, the strength of the confining material is determined by the interaction potential between the molecules, atoms, or ions of the material. Similarly, the cohesive force of the rock that constructs the wall of the cavity determines the capacity of storable energy.

Other cases such as elastic bodies, condensers, coils, sensible heat-storage vessels, *etc.* are similarly treated. However, the mechanical strength of the material depends on temperature. The strength of some materials decreases rapidly with increasing temperature. Even in such cases, this first rule holds at any temperature.

(ii) *The storable amount of energy is independent of the scale of storage structure for a fixed quantity of the structure.*

This rule states that the ratio of the quantity of the storage structure material to the stored energy is constant, being independent on the structure scale, if the maximum energy is stored.

Let us take gas-compression storage as an example. The maximum pressure of the storage gas is assumed to have a constant value. A spherical shape for the storage vessel is most advantageous for a material of constant quantity. The surface stress s (force to enlarge the surface) of the vessel is proportional to the radius of curvature P ($\sigma = kR$), so that the thickness of surface must be increased in proportion to the radius R. Therefore the volume of the wall increases in proportion to R^3 ($4\pi R^2 \sigma$). The storable energy is proportional to the volume ($4\pi R^3/3$), therefore their ratio is constant ($1/3k$).

Similarly, it is understood that the quantity of structural metal of a power storage coil that endures the electromagnetic force is proportional to the storage energy. It should be stressed that a large cylindrical superconducting coil for electrical energy storage has been considered for application in practice. However, it is possible to increase the reserve time constant by scaling up the coil, while the quantity of composed material per storable energy is the same as that of the storage structure in small scale. Use of a superconducting coil has the advantage that

Joulian heat production is zero, while a normal metal coil has the advantage that the ratio of storable energy to Joulian heat production increases with its scale.

How successful energy storage by a superconducting coil will be is decided by the question "How effectively can the strain generated by the electromagnetic force be confined by the storage structure?"

This second rule is also valid in other energy storage systems.

It should be noted that the scale-merit described in Eq.(3.9) signifies an economic saving not the saving of component material.

(iii) *The quantity of component material of the energy storage structure is proportional to the storable energy and the proportional coefficient is constant or nearly constant.*

The storable (maximum) energy in a spherical vessel of compressed gas is expressed by

$$W_m = 2\sigma V/[3(\gamma - 1)], \qquad (4.60)$$

where γ, σ, and V express the ratio of specific heats, stress applied to the material, and volume of component material, respectively. Equation (4.60) is given by assuming an adiabatic compression of ideal gas and the said coefficient is $2/[3(\gamma - 1)]$ ($\gamma = 1.67$ in the case of air compression).

Equation (4.29) can be rewritten as

$$W_m = \frac{1}{2}\sigma V. \qquad (4.61)$$

which expresses the relationship to the volume. The coefficient is (1/2).

We must notice that the coefficient $\frac{1}{2}$ is also valid for the electromagnetic coil.

A general theory of the coefficient is as follows. If the structure for energy storage is elongated in the ratio of $(1 + \epsilon)$ with the maximum stress compared to the state without stress, then the stored elastic energy in the material is

$$W_e = nc\epsilon^2/2, \qquad (4.62)$$

where n expresses the number of elongating directions ($n = 2$ for two-dimensional cases, e.g., spherical vessel, and $n = 1$ for one-dimensional cases, e.g., coil.), c is the corresponding elastic constant.

The stress s is given by

$$\sigma = c\epsilon. \qquad (4.63)$$

When the structure is elongated by ϵ, the stored energy is assumed to change as

$$W_s = W_s^o (1 + \epsilon)^m, \qquad (4.64)$$

where W_s^o is the stored energy in the vessel that has no elongation.

For example, the kinetic energy of ring flywheel is proportional to the reciprocal of the moment of inertia because the angular momentum is conserved. Then $m = 2$ in the case of the ring flywheel. Similarly the magnetic flux is conserved in the case of an electromagnetic coil so that the stored energy is proportional to the reciprocal of the inductance of the coil. The relationship of $m = 2$ also holds for a ring coil.

Another example is energy storage by gas-compression. The stored energy for the adiabatic change is expressed by

$$W_s = W_0 (V/V_0)^{1-\gamma} \qquad (4.65)$$

which is compared to Eqs. (4.60) and (4.64) obtaining $m = 3(\gamma - 1)$.

Using the condition of equilibrium, we have

$$\partial(W_e + W_s)/\partial\epsilon = 0, \qquad (4.66)$$

Table 4.10. Transport of each kind of energy

Energy	Batch	Conveyer, belt, chain	Pipeline	Electric power-line	Optical fibre
mechanical	weight, flywheel, fluid	weight	gas compress, liquid pumping	–	–
electro-magnetic	coil, condensor, electric tray	van de Graff accelerator	–	A.C. D.C.	–
chemical	fuel	fuel	fuel	–	–
photon	–	–	–	–	maser laser
heat	latent & sensible heat (material)	latent & sensible heat (material)	heat pipe (latent heat) sensible heat	–	–

Equations (4.62) and (4.64) are substituted into the above equation, to obtain

$$nc\epsilon V = mV/(1 + \epsilon),$$

If the order of magnitude of $O(\epsilon^2)$ is neglected, then the relationship:

$$W_s = nc\epsilon V/m$$

$$= (n/m)\sigma V \qquad (4.67)$$

holds. This is the third general rule. The coefficient stated in this rule is $(n/m)\sigma$, which does not change appreciably with the structure. Equation (4.67) expresses an important leading principle when an energy storage structure is proposed.

It can be conclusively said that the material from which the structure or vessel is composed is essentially important. The strength of the material at high temperature and high pressure is vital.

4-4. Energy Transport

(1) General survey

The types of energy and transport means are shown in a matrix in Table 4.10. Important transport means are the tanker, pipeline, utility power line, and heat pipe. These facilities have been put into practical use in order to support daily life. Other than these, optical fibers have been widely applied to transport laser light, but the main aim here is not energy transport but signal transmission. The U.S.A., Japan and Europe have been connected using undersea optical cable and important information can now be rapidly exchanged.

The fundamentals of energy transport as well as storage have been already discussed on pages 130 to 133.

The technical terms which have not yet been covered will be explained in the following.

The Van de Graaff accelerator was invented by R.J. van de Graaff (U.S.A.) in 1931. A high tension voltage of 2×10^4 [kV] under atmospheric conditions is provided by this type of accelerator. The principle of van de Graaff's accelerator is shown in Fig.4.19. Positive charge is attached to belt B by a high tension discharging (corona discharging) probe A. The charge is carried by the belt conveyer that is driven by pulleys P_1 and P_2 and given to accumulator K. Conversely, negative charge is carried down by the belt to earth. Continuous transport of electric charge generates, in general, high voltage which can be utilized as a particle accelerator.

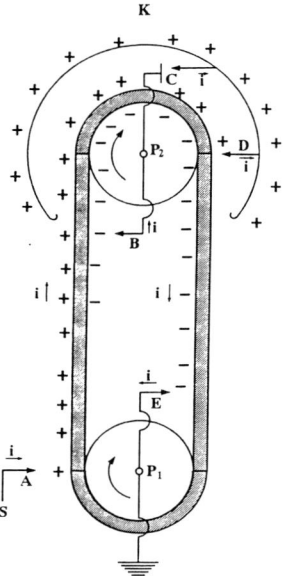

Fig. 4.19. Van de Graaff's accelerator. S: power source, A : for corona discharging, P_1, P_2 : pulley, K : accumulator of positive charge, i : flow of charge.

(2) Costs of transport

Practical and popularized energy transport schemes are classified roughly into (i) batch, (ii) pipeline, and (iii) utility power line. We show the relative costs *vs* distance for some representative energy transport means in Fig.4.20.

The standard is taken as the pipeline transportation of oil. The cost estimation is not exact because it is different, case by case, depending upon time and place. Our new estimation is undertaken by referring to A. L. Hammond *et al.*[19] who state that the cost of electric power transport is reduced by 1/3, because the quality of electric power is 3 times that compared to oil, gas, and coal in caloric equivalence. The reason is that the conversion efficiency from the fuels to electric power is about 1/3.

Some examples are given on Fig.4.20. The cost T of energy transport is evaluated by a similar relationship to Eq.(3.49),

$$T = D + W,$$

where D is depreciation cost of the energy transport facility (tanker, train, pipeline, utility power line *etc.*) and W is the operation cost.

(i) *Transport of oil by pipeline.* The depreciation cost is much higher than the operation cost, and the investment cost per unit distance is independent of the total distance, therefore the transport cost [\$/(kJ·km)] is nearly constant ($D \gg W$). In

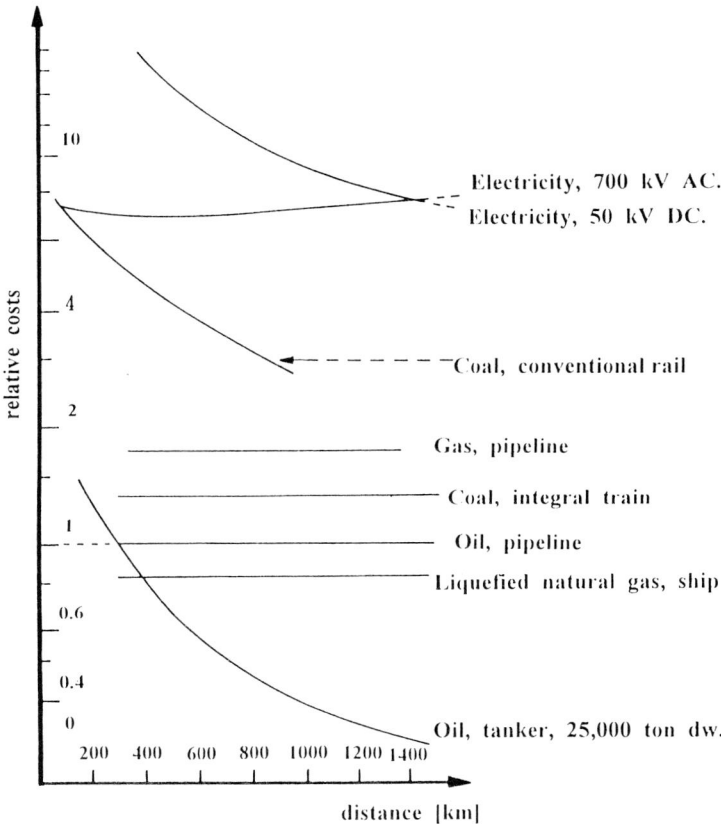

Fig. 4.20. Relative costs *vs* distance curves for energy transportation. The standard is taken as oil transport by pipeline. Transmission of electrical energy is lowered by 1/3 because the conversion efficiency of 1/3 from fuels is taken into account. The electrical energy is three times valuable compares to oil. The curves drawn show only tendency. Taken from A. L. Hammond *et al.*[19].

order to compensate for the pressure drop, the flowing oil is pressurized at pumping stations. This cost is small compared to the investment costs.

(ii) *Gas transport by pipeline.* This is about 1.8 - 2 times more costly compared to that of oil. The reason is simple: the transportable energy density is much smaller than that of oil, in spite of higher volumetric transporting speed ($D > W$).

(iii) *Batch type.* The cheapest energy transport is due to batch type processing such as a tanker where the depreciation cost is also overwhelming ($D \gg W$). Transportation fees increase with distance, while the operational cost does not increase so sharply compared to the fee.

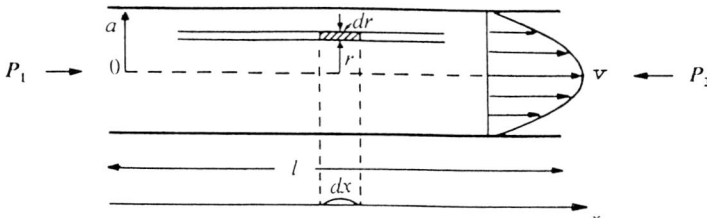

Fig. 4.21. Viscous fluid under pressure difference P_1-P_2.

(iv) *LNG tanker.* Investment cost per transportable calorie of a LNG tanker is much higher than that of an oil tanker because (i) the transportable energy is less, and (ii) the associated heat-insulating equipment is more costly ($D > W$).

(v) *Coal transport.* The reason that the cost of coal transport by conventional railway is high is due to the tariff, imposed by railway companies ($D < W$).

(vi) *Utility power line.* Energy transport by utility power line is most costly. This is due to the costs of facility depreciation and operation being comparatively high ($D \approx W$). We shall discuss electrical power transport later.

(3) Pipeline

Exergy and flow velocity were mentioned in Eqs.(3.15) and (3.16) on page 95 and Eqs.(4.24) and (4.25) on page 132, respectively. Some other important points about pipeline transport are introduced below.

(a) Viscosity. If a nonperfect liquid is flowing with velocity v along the x-axis and a gradient of velocity is generated along the y-axis, then the layer with higher velocity drags the bordering slower layer exerting a shear stress upon the fluid(ref. Fig.4.21). This force is expressed by

$$F = \eta \frac{\partial v}{\partial y}, \qquad (4.68)$$

where the coefficient η is the viscosity whose unit is [Pas] (1 [Pas] = 1[N·s/m^2], and [Pas] means [Pa·s]) and its 10 % value is [poise]. Dynamic viscosity is defined by

$$\eta_d = \eta/\rho, \qquad (4.69)$$

where ρ is the density of fluid. The dynamic viscosity is measured in units of stokes.

$$1[\text{poise}] = 1 \ [\text{dyn·s/cm}^2] = 0.1 \ [\text{Pas}]$$

$$1[\text{stokes}] = 1 \ [\text{cm}^2/\text{s}] = 10^{-4} \ [\text{m}^3/\text{kg}] \ [\text{Pas}] \qquad (4.70)$$

Some examples are shown in Table 4.11. The velocity distribution in a pipeline is also shown in Fig.4.21.

Let us consider a fluid with viscosity η and density ρ flowing in a pipeline with radius a and length l under pressure difference of $P_1 - P_2$. We have the following equation about the rim of a cylindrical part,

$$2\pi l\eta \frac{d}{dr}(r\frac{dv}{dr}) + 2\pi r(P_1 - P_2) = 0, \qquad (4.71)$$

from which the velocity is given by

$$v = (P_1 - P_2)(a^2 - r^2)/(4l\eta), \qquad (4.72)$$

where the boundary conditions:

$$(\frac{dv}{dr})_{r=0} = 0 \quad \text{and} \quad (v)_{r=a} = 0 \qquad (4.73)$$

are assumed.

Table 4.11. Viscosity of fluid at 20°C and 10^3 [hPa]

Gas	10^{-5} Pas	Liquid	10^{-3} Pas	Oil*	Pas
O_2	2.03	H_2O	10.1	Arabian	3.44
H_2	0.88	C_2H_5OH	12.0	light	
CH_4	1.04	C_6H_6	6.5	heavy A	5.33
air	1.8	olive oil	8.4	heavy B	10.8–45
C_2H_4	1.0	machine oil (light)	11.4	heavy C	45–211

*These values are calculated from Stokes values at 50°C.

The transported quantity per unit time is

$$M = \int_0^a 2\pi v r \, dr$$

$$= (\pi a^4/8l\eta)(P_1 - P_2). \quad (4.74)$$

where the units are [m³/s].

Note that the flow rate is proportional to a^4, $(P_1 - P_2)$, $1/l$, and $1/\eta$.

The pressure difference $P_1 - P_2$ is assumed to be entirely used to overcome the viscous resistance, then the energy:

$$\Delta W = \pi r^2 l (P_1 - P_2) \quad (4.75)$$

is removed and converted to heat. For example, we get $\Delta W = 314$ [kJ] by assuming $r = 0.1$ [m], $l = 1,000$ [m], and $P_1 - P_2 = 10^5$ [Pa]. Generating heat in this way is useful to keep the oil's temperature at above a critical value, below which the viscosity of oil becomes so high that flow becomes difficult.

Equations are not so easy to derive in the case of a gas pipe line as a liquid pipeline, because the compressibility of gas exerts complicated effects, and therefore we shall cite the result of an investigation[12].

If a gas with density ρ [kg/m³], viscosity η [Pas] and compressibility k [m²/kg] is fed into a pipeline with diameter a [m] and length l [km] with a pressure of P_1 [kg/m³] and the output pressure is P_2, then the quantity of gas that flows out of the pipeline is, per unit time,

$$M = 130 \left[(P_1^2 - P_2^2)a^5/(\rho k l \eta T)\right]^{1/2}, \quad (4.76)$$

where the unit is [m³/s], and T is the average temperature of gas. Equation (4.76) is an example solution. Some other expressions have been published so far, but there is little difference between them.

(4) Utility power lines

Utility power line systems have been studied for a long time in industrial countries. Control and the associated facilities are also advanced. About three decades ago, the main field of electrical engineering was the control of power line systems. Recently, power transmission lines with one million volt capacities were constructed in industrial countries.

We shall introduce here some essentially important points.

Entropy production by electric current I is

$$\Delta S = I^2/(\sigma T), \tag{4.77}$$

where σ and T are the electrical conductivity and the average temperature of the power line, respectively. Entropy production is due to Joulian heat and proportional the square of electrical current which cannot be avoided unless the resistivity is zero.

If we wish to avoid entropy production by Joulian heat, then transmission by a superconducting wire or microwave is necessary, neither of which has been put into large scale practical use yet. However, the former is utilized on a small scale at laboratories and the latter is widely used in electronic ovens.

Peter E. Glaser (U.S.A.) has proposed a satellite solar power station with 10^7 [kW] of electric power to be transmited to the earth by a microwave system. The input and output antennae need to have diameters as large as 1 [km] and 7 [km], respectively.

Attention must be paid to the global warming accelerated by such power generation. Careful comparison with CO_2 greenhouse effects must be undertaken.

No further mention will be made about superconducting power transmission and microwave systems.

(a) Alternating current transmission. The transmitted power is expressed by

$$P = IV, \tag{4.78}$$

according to which a large P needs larger I or higher V. However entropy production increases with I^2, and therefore it is preferable to increase V. It is, in general, more difficult to generate higher voltages in direct current systems compared to alternating current systems. Therefore A.C. transmission has been the main carrier of current. In addition, a three-phase alternating current system has advantages over two-phase systems, the first being less power loss, the second being fewer connecting lines, and the third being that the instantaneous transmitting power is constant, while the power of two phase A.C. transmission varies with twice the frequency of A.C. current.

A three-phase A.C. power transmission system is now reviewed. Three-phase A.C. power is produced by a generator with triple coils, each of which is placed at a $2\pi/3$ angle in a magnetic field. The electromotive forces V_1, V_2, and V_3 are induced in each coil, and these are given by

$$\left. \begin{array}{l} V_1 = V_0 \sin wt \\ V_2 = V_0 \sin(wt - 2\pi/3) \\ V_3 = V_0 \sin(wt - 4\pi/3) \end{array} \right\} \tag{4.79}$$

The electric currents flowing in each coil are given by, in cases where the load-impedances of these coils are equal.

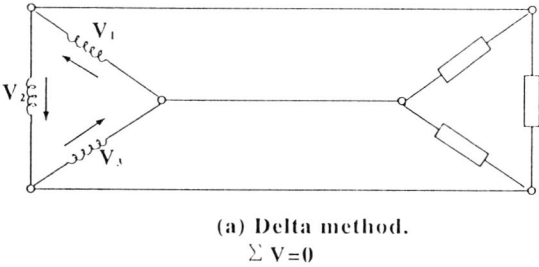

(a) Delta method.
$\sum V = 0$

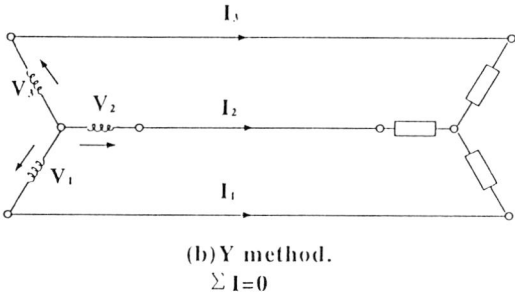

(b) Y method.
$\sum I = 0$

Fig. 4.22. Line connection of three phase alternating currents.

$$\left.\begin{aligned} I_1 &= I_0 \sin(wt - \delta) \\ I_2 &= I_0 \sin(wt - 2\pi/3 - \delta) \\ I_3 &= I_0 \sin(wt - 4\pi/3 - \delta) \end{aligned}\right\} \quad (4.80)$$

The instantaneous P is given from Eqs.(4.79) and (4.80) by

$$P = 3V_0 I_0 \cos \delta. \quad (4.81)$$

The number of utility power lines is not six but only three. We have two kinds of connecting methods. The first is the delta(Δ) method, and another is the Y method. They are shown Figs. 4.22 (a) and (b), respectively. In the case of the delta method, we have $V_1 + V_2 + V_3 = 0$. This equation can easily be derived from Eq.(4.79), so that no current, circulating in the delta loops exists. In the case of the Y method, we similarly have $I_1 + I_2 + I_3 = 0$. Therefore, a fourth line is not needed.

Utility power lines are usually A.C. 200 [V] with three phases and three lines. The voltage is effectively 200 [V] among any pair of the three lines.

The transmitted power averaged over one period τ is calculated as follows.

$$P = \frac{1}{\tau}\int_0^\tau VI dt, \qquad (4.82)$$

into which the first expression for V and I in Eqs(4.79) and (4.80) is substituted. Then we have

$$P = \overline{V}\,\overline{I}\sin\delta$$

$$= (\overline{V}^2/\overline{Z})\sin\delta, \qquad (4.83)$$

where $\overline{V}$ and $\overline{I}$ are the effective values of electric current and voltage, respectively. Z is given by

$$\overline{Z} = V_0/I_0$$

$$= (X^2 + R^2)^{1/2}, \qquad (4.84)$$

and

$$X = 1/(Cw) - Lw, \qquad (4.85)$$

where Z is impedance, X is reactance, C is capacitance, and L is inductance.

In the case of D.C. current transmission, there is no factor with w so that $Z = R$. This means that if the loss by Joulian heat is very small, D.C. power transmission is more advantageous. When the phase difference angle becomes $\pi/2$, P becomes maximum. In order to harmonize the phases, two power generators are operated, one at the forwarding station and another at the receiving station. Adjusting both generators, the difference angle could be $\delta = \pi/6$ at best.

The uppermost capacity of A.C. transmission is ca. 3×10^6 [kW] today. This capacity is sensitive to the climatic conditions.

(b) Direct current transmission.

It was stated in the previous section that, if raising the voltage were easily possible, D.C. current power transmission is more advantageous due to fewer losses and more capacity; this is especially true in long distance transmission. A calculation shows that D.C. transmission is overwhelmingly advantageous at distances over 1,000 [km].

D.C. transmission has become possible since the semiconductor thyristor was invented, which is also called "SCR" (silicon-control-rectifier).

The structure of a thyristor is shown in Fig. 4.23. It has a quadruple structures composed of silicon p-n-p-n junctions. The anode and gate are provided at

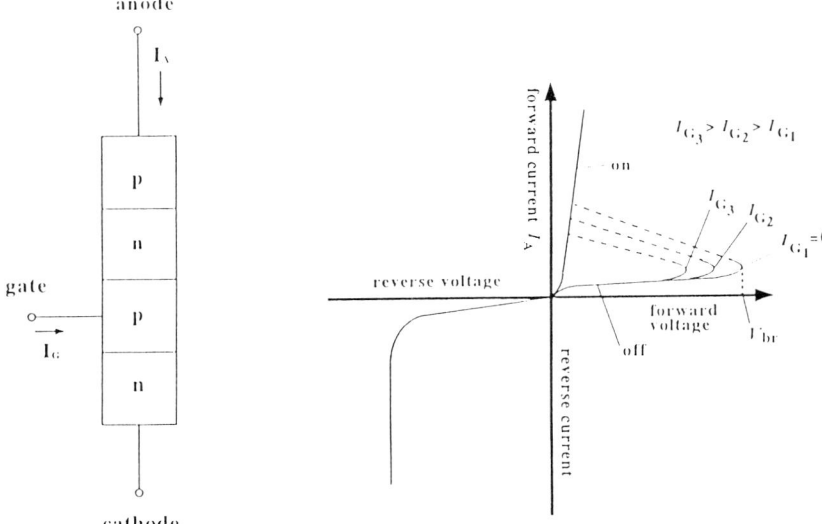

Fig. 4.23. Structure of thyristor.   **Fig. 4.24.** V-I characteristics of thyristor.

the external p-layer and internal p-layer, respectively. If a voltage above a breakover value is applied to the anode, then the forward anode current I_A enters the "on-state". The breakover voltage decreases with increasing gate current (I_G). Once the forward current flows, its control is impossible by the gate and a reverse voltage is necessary.

The current - voltage characteristics of a thyristor are shown in Fig. 4.24. The magnitude of the rectified output can be controlled by adjusting the on-time. The on-time is controlled by an applied A.C. gate current, the phase of which can be adjusted. This is the mechanism of conversion from A.C. to D.C.(**converter**). The reverse conversion of D.C. - A.C. is quite similar and readily understood, *i.e.*, the anode current that has no frequency can be given a phase by an A.C. gate current and the output forward current controlled to have a fixed cycle (**inverter**).

The demerit of D.C. current transmission is that it is impossible to simply take out an optimal lower voltage load from higher voltage transmission lines, while a transformer is readily applied to lower voltage loads in the case of A.C. transmission lines. That is to say, a thyristor system is more complicated and expensive compared to conventional transformer systems.

Chapter 5

Frontier Energy Conversions

PORSHE(Plan of Ocean Raft System for Hydrogen Energy)

Chap. 5. Frontier Energy Conversions

5-1. General Survey

First of all, we shall define "What is frontier?". As has been stated so far, there exist two serious constraints to future energy development; firstly, global-warming as a result of the, CO_2 greenhouse effect, and, secondly apprehensions regarding the depletion of energy resources such as oil and uranium fuels. It is obvious that radical innovations in energy systems and their elementary technologies is required, otherwise the future existence of human beings on the earth will be placed in an extremely precarious position.

Consumption rates and, accordingly, the rates of energy usage have been increasing. One of the important causes is undoubtedly the increase of population in the developing countries. Energy consumption and CO_2 emission rates per capita are increasing in developing countries following the example of the industrial countries. Statistics from 1976 show that energy consumption per capita per day in developing countries is about 10 oil-equivalent-litre/D, below one tenth of that in industrial countries, but since then the consumption rate in developing countries have been sharply increasing, while during the 1990s consumption rates has gradually started to decrease.

World population shares in each area are shown in Fig.5.1, where it can be noted that the population in the industrial countries will be nearly equal to 10 % of that in developing countries in 2060 when oil, gas and uranium fuels are surmised to be depleted. Referring to the tendencies shown in Figs.1,4, 1.5, and 1.6, we will face a rapidly increasing energy requirements and rapidly decreasing production rates not only of oil but also other fuels owing to the spreading effect.

A similar problem occurs with increasing CO_2. The world will be adaptating to environmental damages caused by global warming, a few decades past 2000.

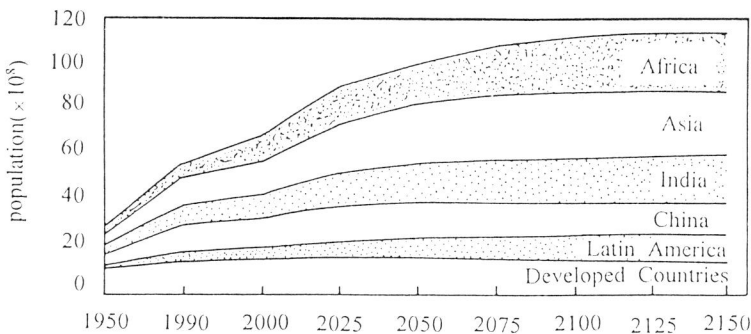

Fig. 5.1. World population in each area (after United Nations).

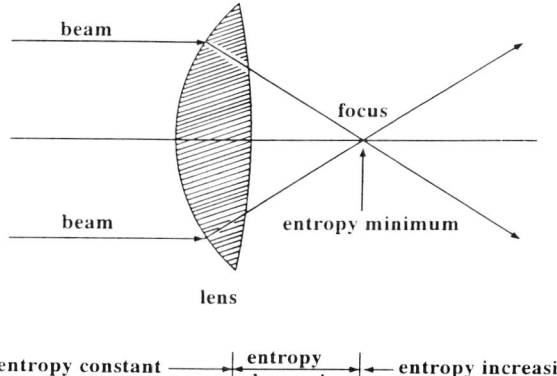

Fig. 5.2. Entropy change in a solar beam collector system.

The definition of "frontier" is thus established as a way of producing energy independent of fossil and nuclear fuels. Answering the question "How is it possible?" will be the definition of the frontier.

The most preferable solutions are summarized in the following and will be the **leading principles of the frontiers**.

(a) Utilization of entropy reduction phenomena. There are innumerable natural phenomena around us. They obey the second rule of thermodynamics which states that the entropy of these phenomena increases. Once the phenomena occur, their available energies are lessened and dispersed. However, if we consider them carefully, we shall find that some component processes of some phenomena possess entropy reducing behavior over a limited space or at the very moment of occurence. Some representative examples are:

(i) *The solar beam collector with a Fresnel lens or concave mirror.* The entropy variation in a solar beam collector system is shown in Fig. 5.2. The amount of incident energy on the surface area of the lens is kept constant at any cross section of a cone whose top and base are the focus and lens area, respectively. On the other hand, the temperature increases up to the focus, therefore the entropy change as shown in Fig. 5.2 is understood.

The highest temperature of a body placed at the focus T_m [K] is determined by the equation:

$$A(T_m - T_o) = (T_m^4 - T_o^4), \tag{5.1}$$

with
$$A = CS\epsilon\, t$$

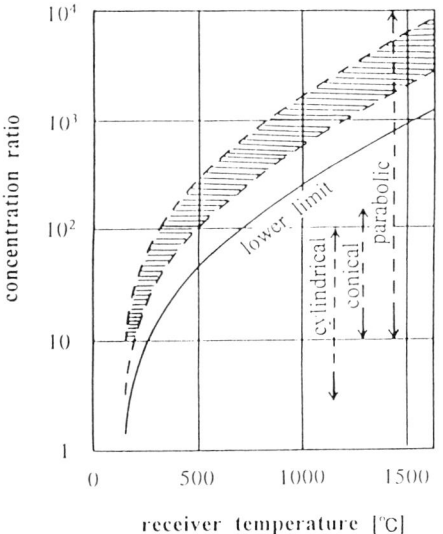

Fig. 5.3. Temperature of the solar collector.

where C, S, ε, and t are the heat capacity, surface area of lens, solar constant, and exposure time, respectively, and the heat losses due to air convection and conduction are neglected.

The saturated temperature T is given by

$$T = (A/\sigma)^{1/3}. \qquad (5.2)$$

The efficiencies of various types of practical collectors vary, and the theoretical saturated temperature *vs* concentration ratio is plotted in Fig.5.3 for some cases.

(ii) *Electric charge separation by solar radiation.* If the solar beam is absorbed into common materials it is converted into, heat at a temperature lower than 150°C in most cases. This means that the entropy increases. However, if the solar photons hit on a semiconductor surface, then pairs of electron-positive holes are generated. Solar cells function by separating the electrons and positive holes by the built-in field of the *p-n* junction, that is, by an effective entropy reduction.

Another chemical method of charge separation is possible. We shall show a wonderful example. The semiconductor TiO_2 (rutile) which is impervious to acid solution, emits two electrons ($2e^-$) and two positive holes ($2p^+$) by absorbing a solar beam. Then each charge reacts on water giving rise to

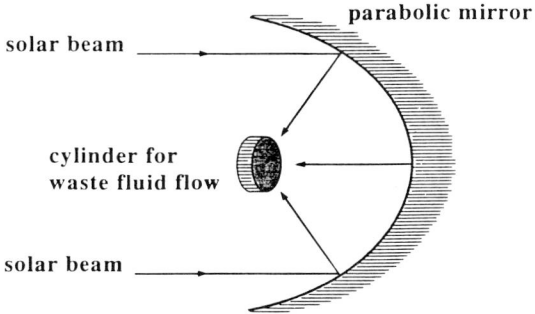

Fig. 5.4. Solar purifier of organic waste fluid.

$$\left.\begin{array}{r} 2e^- + H_2O \rightarrow OH^- + \frac{1}{2}H_2 \\ 2p^+ + H_2O \rightarrow 2H^+ + \frac{1}{2}O_2 \end{array}\right\}. \quad (5.3)$$

Equation (5.3) shows the generation of ion-pairs of OH^- and $2H^+$ alongside water-splitting.

Hydroxyl ions react on any organic materials giving the reaction

$$\text{organic material} + OH^-$$
$$= nH_2O + mCO_2$$
$$+ \text{weak acid}. \quad (5.4)$$

C. Tyner *et al* [*Science* **245**(1992) 130] of Sandia National Laboratories have noticed that this function can be utilized as a purifier of organic liquid wastes. They carried out an experiment as shown in Fig.5.4. Collected solar rays irradiate the TiO_2 particles at the bottom of the sealed tube where waste organic liquid is flowing. When dirty fluid enters the mouth of the tube, clean fluid flows out from its exit.

Solar beam collection has been a traditional frontier energy technology.

(iii) *Other natural processes that reduce entropy, such as the functions of a membrane, catalyst, biological organ, other chemical phenomena, and so on.*

For example, an ion-exchanging separation membrane does work against the Coulomb's attractive force between separated electric charges.

Let the charge and the electric capacity of the membrane be Q and C, respectively, the obtained energy is then

$$\Delta G = Q^2/C$$

174   Energy Technology

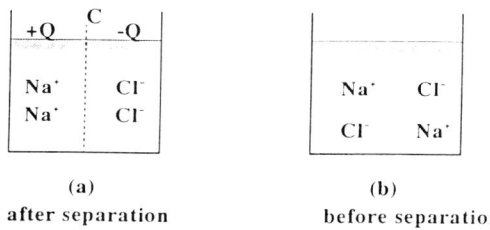

Fig. 5.5. Entropy reduction by an ion-exchange membrane.

No external work is done to the system, so that its change of enthalpy is zero and we have

$$\Delta H = \Delta G + T\Delta S$$
$$= 0,$$

where we have assumed an isothermal reversible change (strictly speaking the temperature will decrease slightly). The work ΔG is positive so that the entropy change is negative ($\Delta S < 0$), *i.e.*,

$$\Delta S = - Q^2/(CT). \qquad (5.5)$$

We also have another entropy reduction process due to the Boltzmann equation for entropy S in relation to thermodynamical probability Ω:

$$S = k \ln \Omega. \qquad (1.30)$$

It is obvious that the probability of finding Na^+ ions in the right half of the cell shown in Fig. 5.5 is 1, while it is $\frac{1}{2}$ in the same region before the separation, so that we have entropy reduction $\Delta S = - R \ln 2$ for one mole of NaCl solution.

The total amount of entropy reduction is given by the sum of Eqs.(5.5) and (1.30). One will be aware that membranes play a notable role in reducing entropy. The functions of a living body are due in great part to its various membranes.

(b) Utilization of entropy minimum phenomena. Next we shall point out some natural phenomena ranked just behind entropy reducing phenomena. This group has entropy production minima. Most entropy reducing phenomena (described in the previous section) are associated with rather static processes but conversely, this group is associated with dynamic phenomena. Examples include superconductivity producing no Joulian heat, heat pipes, cogeneration, *etc.*

We shall comment on heat pipes here, leaving the other topics to be introduced later.

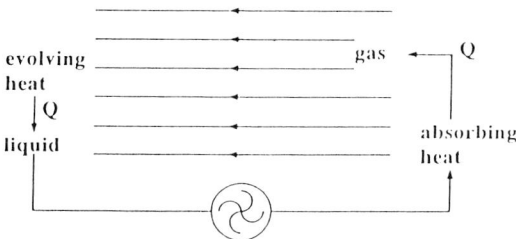

Fig. 5.6. Principle diagram of heat pipe.

Heat transport is one of the most difficult technologies because it is accompanied by entropy production, *i.e.* dispersing heat cannot be avoided (ref. p.188).

In order to transport heat with the smallest loss possible, the latent heat of vaporization should be transported by a force that does not increase entropy or with little entropy production. A pressure difference by one force which the transporting speed can attain the speed of sound. Such transport is diagrammatically shown in Fig. 5.6. The heat transportation rate by heat conduction through a good conductor such as silver or by heat convection cannot compete. The heat medium is gasified by absorbing its heat of vaporization from the heat source and the gas is then transported through a pipe to the low temperature exit by the pressure difference between the entrance and exit. In order to eliminate turbulent flow that could offer resistance to flow, a bundle of thin glass fibres or thin copper wires is inserted into the tube. The transported high temperature vapor is liquified by giving its heat to the thermal load. The volume expansion ratio from liquid to the vaporized gas is on the order of 1,000 so that the pressure difference is automatically generated to drive the carrier gas at speeds as high as the speed of sound. The heat carrier gas is recondensed to liquid which is then pumped to the heat source and so is repeatedly available.

Such an apparatus is called a **heat pipe** and has been developed over many years. The most advanced heat pipes carry heat with a capacity of 100 [W/cm^2] and are conveniently applied to computer-cooling and other electronic equipment cooling. Large-scale heat pipes are also applied to agriculture.

The key technologies are, (1) choice of heat medium, (2) material composing the bundled fibres, (3) material of pipe. The essentially important characteristics of the heat medium are its evaporation and liquefaction temperatures, heat of evaporation, viscosity of vapor, ratio of volume expansion from liquid to gas, non-corrosive property of the medium, and so on. Water is one of the best media, but plasma that is composed of ionized gas could carry more energy (vaporization heat plus ionization energy). An example is introduced later.

Heat storage by metal hydride is also a typical example of a minimum entropy production technology and will be introduced shortly.

(c) Innovation of energy systems. Technical terms of energy systems can be applied to two fields, one of which is the well organized energy convertor having optimum function, and another is the social or industrial circulating systems as mentioned in Chapter 3, where the importance of the effectively functioning hardware in the systems is stressed.

We have mentioned the importance of applying entropy reducing phenomena as well as entropy production minimum phenomena; however, it will be more important to combine both phenomena to construct an ideal conversion system.

We shall point out the importance of a cooperative arrangement for refrigeration between electric power and city gas systems during summer in urban areas. Refrigeration in summer is usually done by electric power, therefore the maximum capacity of power generation of a country can hardly provide the total power demand sometimes. More than 90 % of the total generation wattage capacity were consumed in the 1990 summer in Japan. Conversely, city gas during that period is usually not utilized. City gas is an excellent energy source that can also be converted to refrigerating capacity. However, no such cooperation is established in most countries.

For this reason we will survey refrigeration methods later and methods for applying gas energy to them.

5-2. Entropy Reduction Systems

Entropy reduction methods for creating available energy (exergy) from existing dispersive energy sources are discussed in this section. However, complex systems, composed of multistage processes, produce entropy which is slightly different from a simple system such as collection of solar energy. The water-splitting system by photoelectrode, or solar energy storage systems using photoelectrochemical reactions in the following belong to this complex type.

(1) Water-splitting by semiconductor electrode

Photon energy can be converted to electrical energy as described in Chapter 2 (p.80), and is also available to equipment such as a purifier for waste liquid (p.173). We shall now introduce another phenomenon which uses a semiconductor electrode to split water.

A simple arrangement of the experimental apparatus is shown in Fig.5.7. Ultra-violet light radiates on the TiO_2 semiconductor surface (1), then two pairs of electrons (e^-) and positive holes (p^+) are created:

$$(TiO_2) + h\nu \rightarrow 2e^- + 2p^+$$

At the semiconductor electrode called the photo-semiconductor electrode, the reaction shown in the second equation of (5.3) occurs.

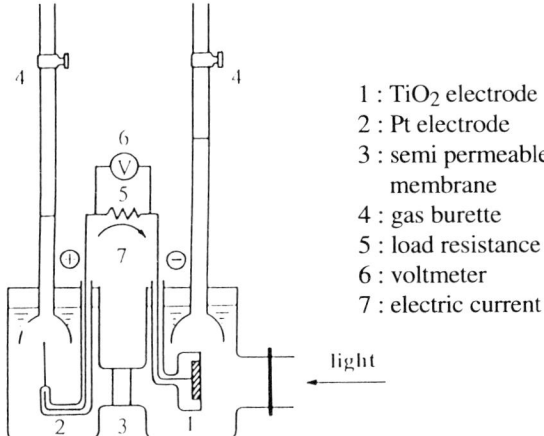

Fig. 5.7. Water-splitting by the photoelectrode method.

$$2p^+ + H_2O \rightarrow 2H^+ + \frac{1}{2}O_2.$$

The two protons ($2p^+$) migrate to the counter electrode (2, Pt) through a semi-permeable membrane (3, Agar bridge) placed in the electrolyte (*e.g.*, KOH 30 %). At the Pt electrode the reaction shown in the first equation of (5.3) occurs

$$2e^- + H_2O \rightarrow OH^- + \frac{1}{2}H_2.$$

The two electrons ($2e^-$) generated at the Pt-electrode are led to a photoelectrode *via* a load resistance (5) to combine the two protons to form water. This system splits water by incident light. K. Honda and A. Fujishima[4,17] of Tokyo University reported this phenomenon as early as 1969 and it has attracted great interest since then.

The free energies existing in the natural world are chemical energy, almost all of which is contained in fossil fuels, and solar photon energy. If the latter can be converted to hydrogen evolved from water, then hydrogen can replace electrical power. Electrical energy is difficult to store, while hydrogen is easily stored. Hydrogen burns to generate only water, that is, it is as clean as electrical energy. A system that utilizes hydrogen as secondary energy is called a hydrogen energy system and was introduced on p.130 (Fig.4.8).

Water-splitting will be discussed in detail in a later section. We have some typical water-splitting methods such as electrolysis (p.149) and the photochemical or thermochemical cycle method (p.82, p.202), but it is more important to utilize solar energy by applying it to the "photolysis of water" (which means direct water-splitting by photons).

178   Energy Technology

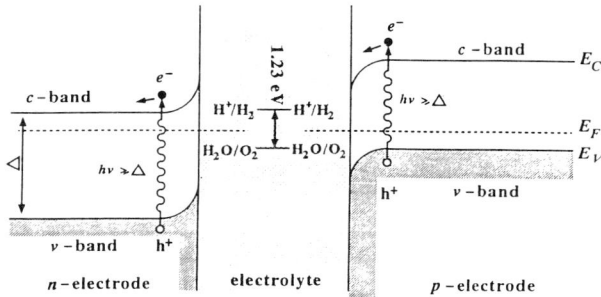

Fig. 5.8 Energy band structure at the boundary of n- and p-electrodes in an electrolyte.

Let us study why water is split in such an experiment, using Fig.5.8. Notice that the boundary between the *n*-type electrode and the electrolyte, and the bottom of the conduction band (*c*-band) and the top of the valence band (*v*-band), are bent up at the boundary because of the resulting space charge. Due to this built-in field, the electrons in the *c*-band escape from the boundary to the counter electrode. Conversely, positive holes in the *v*-band get together at the boundary. Therefore the *n*-type electrode is positively charged, and counter electrode becomes negative.

If the potential energy difference between the separated electrons and positive holes is larger than the necessary electrolysis voltage (1.22 [V] + over voltage), then water can be split by the said chemical reactions. In the right half of the figure, the energy diagram at the boundary between the *p*-type electrode and electrolyte is shown. Similar separation of charges occurs so that the *p*-type electrode becomes negative, while the *n*-type electrode becomes positive.

If the Pt electrode is replaced by a *p*-type semiconductor, water-splitting occurs more readily. Comparing this mechanism of water-splitting with solar cell generation combined with water electrolysis, it can be seen that both methods are equivalent. Both processes reduce the entropy of the system. However a solar cell combined with water electrolysis is a rather complicated system compared to the photoelectrode system.

Next we shall consider the properties of semiconductors necessary for use as photoelectrodes.

(i) *High quantum efficiency.* The number ratio of generated electron-positive hole pairs to incident photons should be large. The energy gap between the top of the *v*-band and the bottom of the *c*-band must be a little smaller than the incident energy of light. The quantum yield depends on the characteristics of the semiconductors.

(ii) *Sufficient absorption of incident beam by the semiconductor is needed.* Some coating for preventing light reflection is needed. This coating should not be dissolved in the alkali electrolyte.

(iii) *The photoelectrode must be stable in the electrolyte.* Semiconductors in this process are exposed to alkali or acid electrolytes to which most of them are not

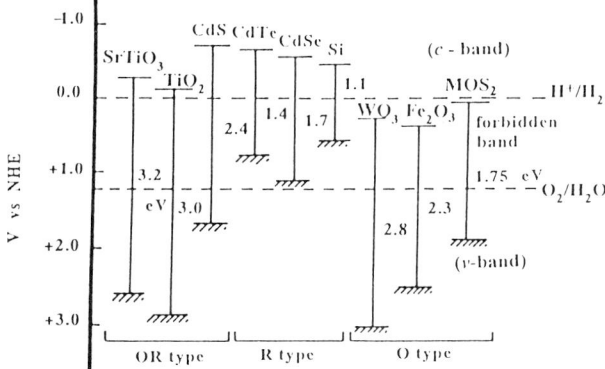

Fig. 5.9. Relative energy levels of semiconductors in electrolyte with pH=0. NHE : normal hydrogen electrode, O : oxidation, R : reduction

chemically stable. Some semiconductors react with electrolytes when they are irradiated by light even if they are stable in the dark. This phenomenon is called "photoerosion".

(**iv**) *Large output of photovoltage based on the large band bending of semiconductor and good reversibility of redox (abbreviation of reduction and oxidation) reactions in the electrolyte.* It is nearly impossible to predict the magnitude of the bending due to space charges in the semiconductor. Generally speaking, multicomponent semiconducting alloys have large space charge effects.

(**v**) *No filtering effect of the electrolyte is required.* It is noticed that use of non-poisonous redox species requires little filtering effect.

The relative positions of the top of the *v*-band and the bottom of the *c*-band are shown in Fig. 5.9 for some semiconductors that have been studied so far. The vertical axis of the figure is the voltage relative to the normal hydrogen electrode (NHE) and for a solution with pH = 0. The voltage between H^+/H_2 and O_2/H_2O (the second reaction formula in this section) is 1.23 [V]. One should note that semiconductors such as CdS, CdTe, CdSe, and shown in the figure are subject to erosion and are therefore not available for use. If they are used, then some regenerative substance should be mixed into the electrolyte solution to recover the eroded materials.

In order to utilize solar photons, the width of the energy gap must be matched to the average photon energy, *i.e.*, *ca.* 1 [eV].

The efficiency of water-splitting by photon energy is defined by

$$\eta_p = (\text{LHV of evolved } H_2) / (\text{incident photon energy})$$

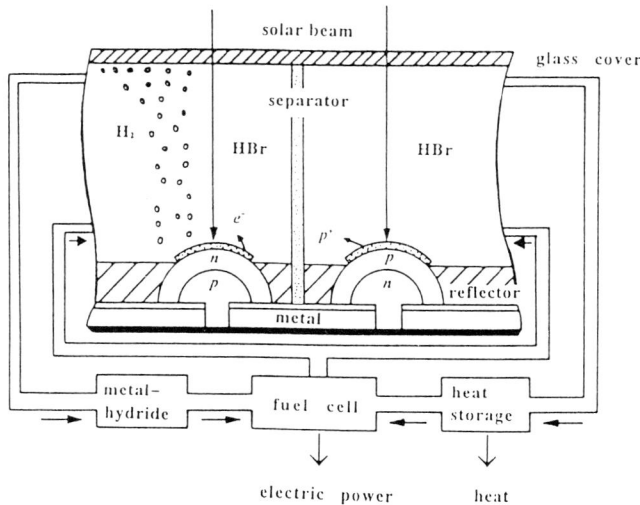

Fig. 5.10. Design of solar battery after Posa[41].

the value of which is not necessarily large. However, one must pay attention to the fact that hydrogen can be stably stored, while electrical power cannot, and hydrogen can be used as a basic chemical material.

(2) Solar battery

Solar cells are able to generate electric power but unable to store it. We shall now introduce an idea by Texas Instruments Inc. (U.S.A.)[14,41], which may be called a "solar battery" because the generated electric power can be stored on site and regenerated whenever required. The principle of this solar battery is shown in Fig. 5.10.

Solar beams enter a cell composed of two chambers that are filled with an aqueous solution of hydrogen bromide (HBr) and impinge on spherical electrodes fixed at the bottom of the cell. Two kinds of solar cells, n on p- type and p on n-type, are placed in the cathode and anode chambers, respectively. The solar beams entering the cathode chamber hit the n on p solar cell, and electrons are emitted into the HBr solution and reacts on HBr to evolve $1/2 H_2$, *i.e.*,

$$H^+ + e^- = \frac{1}{2} H_2.$$

The evolved hydrogen is led to metal hydride and stored.

The solar beams entering the anode chamber turn out positive holes from the p on n solar cells and the positive holes cause the reaction :

$$Br^- + p^+ = \frac{1}{2} Br_2$$

The evolved bromine is sent to a storage vessel where its heat is extracted by a heat exchanger. The bromine is liquefied and stored. Whenever electrical power is needed, a fuel cell is fed by hydrogen and bromine gases to generate power which has a voltage of 0.258 [V].

The exhausted gas from the fuel cell is HBr, which is recovered and put into the battery cell.

This conception has some difficulties. First the Si solar cells are eroded by the HBr acid unless a coating is applied. The solar beam is then doubly weakened by the coating and the HBr solution. Next, the HBr fuel cell has not been investigated yet and an erosive gas like Br_2 is not easy to handle. These trial experiments seem to be far from practical use at the present stage.

(3) Membrane

Let us introduce the phenomenon of separation by a membrane. Inserting a membrane that separates a container, where a solute A and another solute B aredissolved in a solvent, into two parts, then three possible states of separation occur. The first is when only solute A can penetrate the membrane, the second that solute A is replaced by B, and the third is that only the solvent can pass through the membrane. Both the first and the second are called "**dialysis**" and the third is called"**osmosis**". In order to give rise to separative phenomena, an energy to generate the driving force is necessary.

Table 5.1. Membrane separation and the corresponding energy

Intensive variable (generalized force)	Energy	Osmosis	Dialysis
pressure (osmosis pressure)	mechanical	osmosis, reverse-osmosis, penetration	piezo-dialysis
thermal gradient	heat	thermal osmosis	thermal dialysis
electrical potential	electrical	electrical osmosis	electrical dialysis
density gradient	chemical	osmosis	dialysis, gas-dialysis
chemical potential	chemical	chemical osmosis	chemical dialysis
photochemical potential *	photon	photon-osmosis	photon-dialysis

* The technical term "photochemical potential" is not formally decided yet.

A classification of the types of membrane separation is given in Table 5.1, where the driving forces (intensive variables), the corresponding energies, osmosis and dialysis are set in order. Energies are classified into five categories, according to which the membrane separations are also classified.

These separations are basically due to the following processes[27].

(i) *Separation is a result of the differences in mean free path, or molecular or ionic size.* In most gas separation, the mean free path is smaller than the pore or the small cavity of the membrane and hence can pass through the membrane. Osmosis or dialysis is sometimes realized in such simple a way.

Molecular mean free path is given by

$$\lambda = 1/(\sqrt{2}\pi N d^2) \quad (5.6)$$

or

$$\lambda = (kT)/(\pi P d^2) \quad (5.7)$$

where N, d and P are the molecular density, molecular diameter, and gas pressure, respectively.

The molecular diameter and the mean free path of some commom molecules are shown in Table 5.2. It is easier to separate molecules having smaller diameters and mean free paths. It is noticed that the molecules with smaller diameters have, in general, longer mean free paths.

Equations (5.6) and (5.7) show that the density gradient, thermal gradient, and pressure gradient give rise to the effective separation rate. It should be noticed from this table that CO_2 gas is not so easy to separate from N_2 or H_2O by this method. Separation of CO_2 gas is undertaken by another method.

Table 5.2. Diameter and mean free path of some gas molecules (after *Smithsonian Physical Tables*, 8th Ed.)

Molecule	d [nm]	λ [nm]
He	19.0	331.0
H_2	24.0	154.0
O_2	29.8	87.8
N_2	31.5	82.1
CO_2	33.4	54.4
H_2O	33.3	54.5

(ii) *The material used as a membrane can chemically mediate the molecules, atoms, and ions in passing these particles through the membrane.* This is an essentially important mechanism in most cases of practical separators. It is not always true that solute particles with smaller diameters are easier to pass through the membrane. As a matter of fact, membranes exist that pass only the Cl$^-$ ion in the dialysis of salt water (NaCl), although the diameter of the Na$^+$ ion (9.5 [nm]) is much larger than that of the Cl$^-$ ion (18.1 [nm]). This is due to a chemical process where the Cl$^-$ ion has a much stronger potential for combining with the membrane material. The Cl$^-$ ions that enter into the membrane can migrate toward the other side and leave the membrane.

(iii) *We shall now give brief comments on osmosis and dialysis in Table 5.1.*

Piezo-osmosis or dialysis is mainly a result of the principle of "higher pressure and smaller diameter of solute particle" which tends to increase the collision frequency of the particles on the surface of the membrane.

In the case of thermal osmosis or dialysis, the chemical reaction between the solute particle and the membrane material is thermally enhanced on the higher temperature side. Similar mechanisms are realized in photon osmosis or dialysis, that is, the irradiation of light enhances the chemical reaction between the solute-ion and the membrane material. Electrical osmosis and dialysis are readily understood. Not only ions but also electrons are driven through the solution and in the membrane, respectively. Therefore, a prominent separative effect is expected.

Density gradient osmosis and dialysis were already introduced on p. 9 (ref. Fig.1.1) and are the most popular phenomena. The density gradient force (which is one of the generalized forces) is a typical intensive variable whose name is chemical potential.

(iv) *Salt and fresh water electric power generation.* Here, we introduce an experiment on electric power generation using an ion-exchange membrane. A preliminary outline of this experiment was shown in Fig.1.1. The following describes it in more detail.

A fresh water flow is maintained in a cell that lies between two other cells. These two cells are filled with salt water. The wall between the middle and both side cells is separated by cation- and anion-exchange membranes. The structure of the three cells is shown in Fig.5.11.

A polymer acid, composed of synthesized resin (R) combined with hydroxyl groups, sulfate, or other acid salts (AH) is called an anion exchange membrane (anion is the name hence it is generated, having negative charge).

On both sides of the anion exchange membrane, fresh water penetrates into the membrane and the following reaction occurs

$$RAH + H_2O \rightarrow RA^- + H_3O^+ \qquad (5.8)$$

where the RA$^-$ ion reacts on the cation Na$^+$ of the salt water. Then we have

$$RA^- + Na^+ \rightarrow RANa, \qquad (5.9)$$

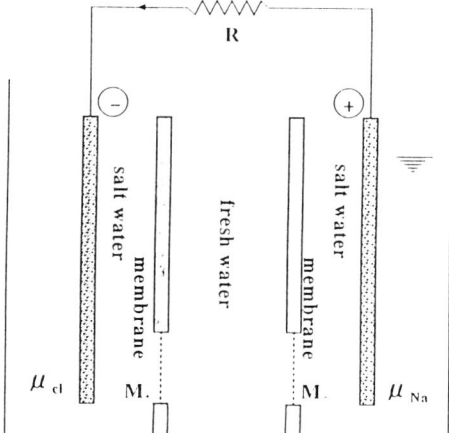

Fig. 5.11. Principle of fresh-salt water power generation. An application of Fig.1.1 is shown for river fresh water *vs* salt sea water power generation. The cell is composed of three compartments, in the middle of which the fresh water flows and the salt water flows in the both sides. The membrane M_ passes only Cl⁻ ions into the middle room so that the right hand side room becomes Na⁺- rich to be positively charged. The similar process occurs in the left room and becomes negatively charged.

where RANa is not dissolved in water. Therefore only Na^+ ions in the salt water are removed so that the remained Cl^- density increases. As the result, the cell room which is separated by anion exchange membrane is negatively charged. The Na^+ ions are emitted out to the central cell.

Due to a similar mechanism, the Cl^- ions removed from the cell that is separated by a cation exchange membrane becomes positively charged.

Thus we have electrical generation whose voltage is given by

$$V = (2f/F)(\mu_1 - \mu_2), \tag{1.27}$$

where μ_{ij} are the relative chemical potentials in both side cells.

Now we show an experimental result. Sea water near the mouth of a river and the fresh water in a river are assumed to have concentrations of 0.45 [N] and 0.125 [N] at 20°C, respectively. Membrane thickness, membrane area, and the membrane interval are measured to be 0.1 [mm], 1 [m²], and 0.1 [mm], respectively. The internal electrical resistance of both side cells, the middle cell, the cation exchange membrane, and the anion exchange membrane, are 0.28 [mΩ], 0.82 [mΩ], and 0.03 [mΩ], respectively. The total resistivity is 1.15 [mΩ]. The measured voltage is 0.06 [V], while the theoretical voltage is 0.066 [V].

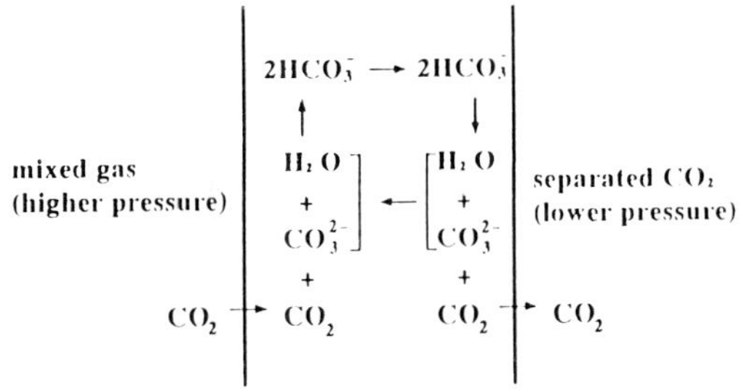

```
                    2HCO₃⁻  →  2HCO₃⁻
                       ↑             ↓
 mixed gas          ⎡ H₂O ⎤     ⎡ H₂O ⎤        separated CO₂
(higher pressure)   ⎢  +  ⎥  ←  ⎢  +  ⎥        (lower pressure)
                    ⎣ CO₃²⁻⎦    ⎣ CO₃²⁻⎦
                       +             +
              CO₂ ⎯⎯⎯ CO₂          CO₂ ⎯⎯⎯ CO₂
```

$$CO_2 + H_2O + CO_3^{2-} \rightleftharpoons 2HCO_3^-$$

Fig. 5.12. Aqueous solution of carbonate transport mechanism of CO_2 in aqueous solution of carbonate.

If the load resistance is 1.15 [mΩ], the same as the internal resistance, then the current is 26 [A] and the wattage is 1.56 [W].

In order to realize this system of salt-fresh water power generation, the followings must be necessary. 1. The fresh water must be continuously renewed. The driving force would be wave or river water flow energy. 2. The river and the sea water must be kept clean. 3. The generated electric power has a low voltage and direct current which is conveniently applied to water-electrolysis to produce hydrogen.

(v) *Separation of CO_2 by membrane.* Recovering CO_2 from the exhausted gas of fossil fuel combustion is an important technology. Cellulose acetate membranes are able to separate CO_2 gases. However, water molecules are not separated from CO_2.

Another separation method is due to liquid membrane. If the solubility of CO_2 gas into the liquid membrane is strong and the reactitvity between the CO_2 molecule and the carrier molecule in the liquid membrane is also high, the absorbed CO_2 molecules are effectively transported to opposite side of the membrane so that CO_2 is separated. Piezo-osmosis as an example of such cases is diagramatically shown in Fig.5.12.

(4) Catalyst

Coexisting with reactants, catalysts (catalyzers) enhance the chemical reaction, but is stoichiometrically independent of the reaction. Therefore the catalyst does not appear in the equation of chemical reaction. F.W. Ostwald (1853 - 1932) defined that catalysts exert no effect on the reaction equilibrium.

Chemical industries today utilize catalysts very commmonly. Catalysts are used in more than 80 % of industrial chemical processes.

Three basic processes of catalysis are absorption of reactant into the catalyst, chemical reaction at the catalyst surface, and the liberation of the chemical products.

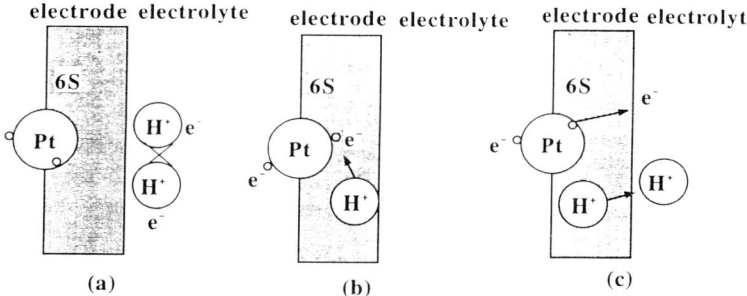

Fig. 5.13. Catalysis of platinum at the H_2 - feeding cathode in a fuel cell.

The electrons in the outermost orbit of the electronic shell play an important role in the case of solid catalysts such as transition metals and their complexes.

Another catalysis function is represented by Zeolite which is one of the most commomly used catalysts and will be introduced later. Zeolite that can be artificially manufactured is porous and easily absorbs chemical reactants into itself.

(i) *Platinum catalysis.* We shall explain the qualitative mechanism of platinum (Pt) catalysts in the case of an H_2 - air fuel cell (p.144).

The catalysis mechanism at the interface between the Pt electrode and the electrolyte is shown in Fig.5.13, where electron transfer from the input hydrogen molecule to the electrode metal is pictured. An outline of the process has already been mentioned (p.144).

The following reaction occurs at the cell cathode,

$$H_2 \rightarrow 2H(a), \quad 2H(a) \rightarrow 2H^+ + 2e^-$$

where (a) represents the adsorbed atom. The released electron goes to the opposite electrode *via* an external load. At the oxygen fed anode, the reaction

$$O_2 \rightarrow 2O(a),$$

$$2O(a) + H_2 + 2e^- \rightarrow 2OH^-(a)$$

is obtained where a catalyst is also needed. The generated $2OH^-$ reacts on $2H^+$ at the boundary membrane and water, $2H_2O$, is yielded.

There is only one electron in the outermost 6S electronic shell of Pt, while the capacity of the S-shell is two electrons. A hydrogen molecule is composed of $2H^+$ and $2e^-$ that are commonly shared by the two protons (Fig.5.13, a). One of the electrons transfer from H_2 to the vacant position of the Pt orbit, because it is more stable (Fig.5.13, b). The third step is that one of the two electrons in the outermost orbit of Pt is transferred to the metal cathode, because the cathode is biased. The remaining proton is released into the electrolyte(Fig.5.13, c)

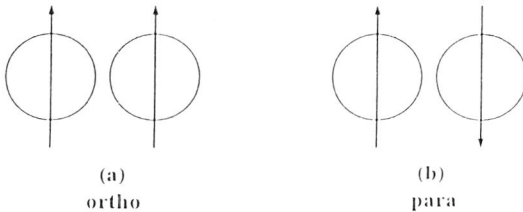

(a) ortho   (b) para

Fig. 5.14. Ortho - and para - hydrogen.

(ii) *Para-ortho conversion.* A hydrogen molecule is composed of two hydrogen atoms. If the nuclear spin of each atom is parallel to the other, then the molecule is called "ortho-hydrogen". On the other hand, hydrogen whose two nuclear spins are anti-parallel is called "para-hydrogen". Figure 5.14 shows the difference between these two kinds.

Para-hydrogen is more stable than ortho-hydrogen and the energy difference between them is about 744 [J/g] at 20 [K] and this is the so-called the conversion heat. The equilibrium ratio of ortho-hydrogen to para-hydrogen is about three to one at room temperature, but it decreases with decreasing temperature. When ortho-hydrogen is converted to para-hydrogen, conversion heat is evolved. Therefore a heat loss is realized in the cryogenic storage of hydrogen. To avoid the loss, ortho-para conversion is necessary before liquefaction. This is undertaken in the presence of ferrous hydroxide $Fe(OH)_2$ catalyst. Catalysis in this case is spin-spin interaction between hydrogen and the catalyst. The ortho-para conversion heat is shown in Fig.5.15.

(iii) *Zeolite.* Zeolite is a famous catalyst widely utilized in chemical industries such as oil refineries. Zeolite is manufactured not only from basic volcanic rock but also by synthesis. Its chemical formula is

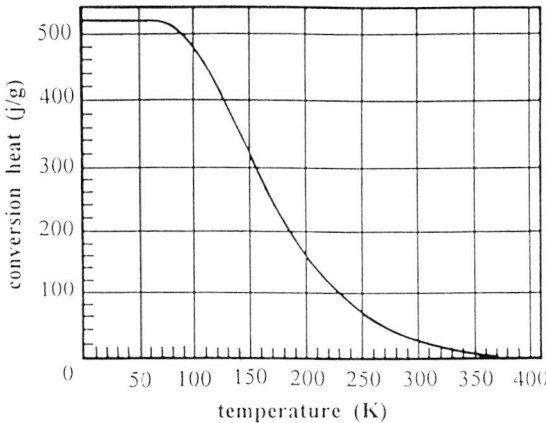

Fig. 5.15. Conversion heat from ortho- to para-hydrogen.

$$M_{2n}O \cdot Al_2O_3 \cdot xSiO_2 \cdot yH_2O \tag{5.10}$$

which shows that Zeolite is basically a compound of alumina (Al_2O_3) and silicic acid ($xSiO_2 \cdot yH_2O$). The notation in (5.10) is as follows: M = Na, K, Ca, Ba, x = 2 - 10, y = 2 - 7, and n is the valence number.

The crystal structure is characterized by an aluminosilicate tetrahedra frame work. This structure and the ordered array of molecules form pores, or windows, opening into cavities into which ions, water molecules, and other chemicals can fit. Due to these properties, ion-exchangeable large cations, and loosely held water molecules permit reversible dehydration.

Zeolite is used not only as a catalyst but also as a molecular sieve, which is able to separate molecules, adsorption material, water softener, *etc.*.

The detailed behavior of Zeolite catalysis has not yet been perfectly studied and patent disputes about its manufacture never cease.

5-3. Entropy Production Minimum Systems

Not many natural phenomena with entropy reduction exist. However, one may expect a considerable number of entropy minimum phenomena under certain conditions.

However, we shall mainly introduce systems whose entropy production is relatively low in this section, and describe some typical examples.

(1) Functionality materials

Zeolite and other catalysts can be undoubtedly classified as functionality materials. The examples mentioned in this section are chosen because their dynamic characteristics are clear and uncommonly precious. Not only (i) *AMTEC* but also **(2) Metal hydrides** and **(3) Superconductors** are functional material.

(i) *AMTEC (alkali-metal thermoelectric converter).* As has been mentioned already (p.175), heat transport by a "jet" of vaporization heat is one of the entropy production minimum phenomena. The vapor jet is replaced by "plasma jet" if some functionality material is applied. AMTEC is a typical example. This is a type of thermoelectric converter discussed in the previous chapter.

Electric charge transport in the thermoelectric converter is associated entirely with the heat conduction processes which produce a considerable amount of entropy in the system. The driving force (thermal gradient) is replaced by pressure in the AMTEC process, so that the entropy production is much reduced.

The alkali metal of AMTEC is beta-alumina (β''-alumina) whose chemical formula is

$$Na_5LiAl_{32}O_{51} \tag{5.11}$$

Very few atoms of sodium and lithium are alloyed in alumina. This kind of materials is called solid electrolyte and is sometimes used in high temperature

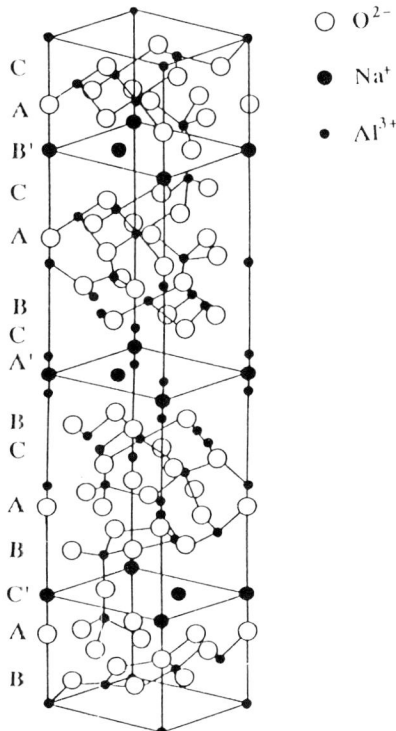

Fig. 5.16. Crystal structure of β"-alumina.

lectrolysis. The structure of β"-alumina is complicated and is shown in Fig.5.16.

Two kinds of ions, Na^+ and O_2^-, are sparsely situated and 1/6 of the positions for Na^+-ions are vacant in the atomic planes denoted by A', B' and C'. Therefore the Na^+-ion can migrate freely throughout these atomic planes. Electrical resistivity at 1,000 [K] is only 1.4 [Ωcm], which is of the same order of magnitude as semiconductors. Applying this property of β"-alumina, a plasma flow of Na^+-vapor is actuated by pressure in this solid electrolyte, and thermoelectric conversion is possible.

A system diagram for AMTEC is shown in Fig.5.17[10]. The Na^+ gas is vaporized from liquid metal sodium that is heated to temperatures as high as 1,200 [K] and enters the β"-alumina, being subjected to a strong pressure produced by the high vapor pressure of sodium. This is vaporization from liquid sodium to the β"-alumina.

The Na^+-ions migrate through the atomic planes and when they reach the porous electrode they catch the electrons that come *via* an external load. These

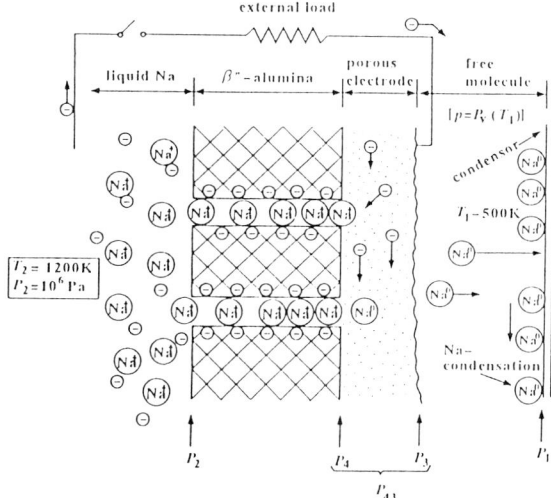

Fig. 5.17. System diagram of AMTEC after T. Cole[10].

electrons are generated from Na atoms (Na → Na$^+$ + e$^-$) at high temperature. The Na$^+$-ions combine with the electrons and the Na atoms are recovered (Na$^+$ + e$^-$ → Na).

This neutral sodium vapor is recondensed to become liquid. The liquid is sent back to the heater by an electromagnetic pump.

The output voltage is determined by the pressure ratio between the sodium vapor at the higher temperature side and the porous electrode and assumes values between 0.4 and 0.1 [V] depending on the ratio.

Heat is transported, accompanied with the Na$^+$ ions, that is, the Na atoms absorb heat when vaporized and exhaust it when condensed. The heat transport is the same as in the case of a heat pipe. The coupling constant between the heat current and electric current is very high compared to the semiconductor thermoelectric converter, where more than half of the heat is transported by a phonon system. The temperature difference between both sides of the β"-alumina, the entrance and exit of Na$^+$-ion, is very small, having small heat loss. The driving force on the Na$^+$ ions is not the thermal gradient but the pressure, so that the unavailable heat is very low.

Of primary importance in AMTEC is the choice of alkali metal. Materials must fit the crystal structure and be stable at high temperature. Next is the system design. We show in Fig.5.18 a new type of AMTEC, where two kinds of wick are utilized. The Na vapor is ionized passing through the second wick[47].

The conversion efficiency of AMTEC is given by

$$\eta = IV / [I(V + [L + C_p \Delta T]/F) + Q], \qquad (5.12)$$

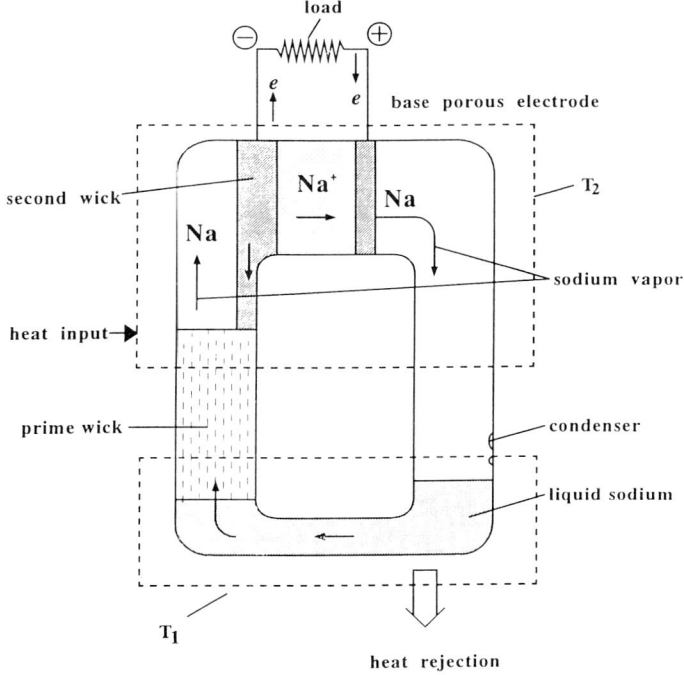

Fig. 5.18. Wick-return AMTEC after K. Tanaka et al. (47)

where IV, L, C_p, ΔT, F, and Q are the electrical work at the load, heat of vaporization of liquid sodium, heat capacity at constant pressure, the temperature difference, Faraday's constant, and the dispersed heat of the system, respectively. Assuming the conditions : $T = 850$ [K] ($T_2 = 1{,}200$ [K]) and $Q = 0$, then we get $\eta = 35$ % and the current density is as large as 2.6 [A/cm^2]. This value is about ten times larger than a conventional thermoelectric generator.

(2) Metal hydride

Hydrogen embrittlement is a well known-phenomenon and is a source of failure in ferrous materials. This phenomenon is explained by the fact that hydrogen atoms or ions easily enter the lattice interstices and combine with each other or with carbon atoms becoming hydrogen molecules or methane molecules. The gas generated from these reactions gives rise to large pressures in the lattice and destruction results.

Contrary to the hydrogen embrittlement, some metals and alloys can absorb hydrogen atoms in their interior body to as much as 700 to 1,000 times of their volume. This phenomenon was discovered by T. Graham (1805 - 1869) more than 100 years ago.

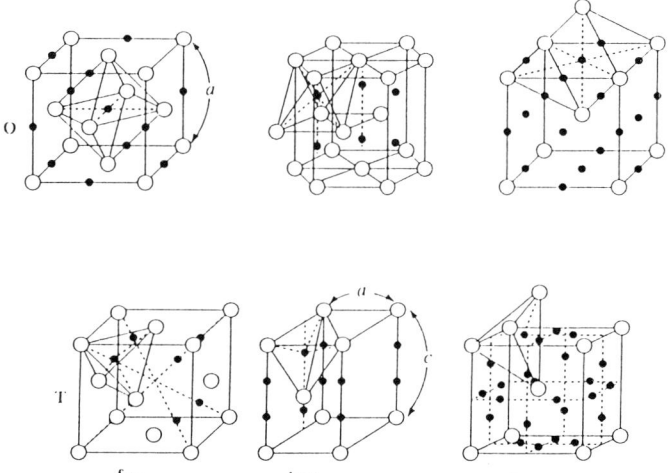

Fig. 5.19. Hydrogen atoms in metal lattice.
○ : Metal Atom, ● : Hydrogen Atom

In conventional energy systems, protection of containers, pipelines, and other equipment from hydrogen embrittlement has been a vital task. However, the next important task will be to develop a new system to utilize hydrogen storage by metals or alloys.

(i) *General survey.* Metals or alloys (M) that absorb hydrogen and dissociate it reversibly are called "metal hydrides". The reaction of a metal - metal hydride is

$$H_2 + 2M \rightarrow 2MH + Q, \tag{5.13}$$

where Q is the reaction heat, 17 - 20 [kJ/mol-H_2] in most cases.

Reacting hydrogen on alkali or alkali-earth metals, ionic crystals are generated. Covalent materials are obtained when hydrogen reacts on silicon or aluminum. A stoichiometric hydrogen number is required when using the materials described so far, but no stoichiometric condition is imposed in the case of the absorption reaction in Eq.(5.13).

Transition metals such as Ti, V, Cr, Mn, Fe, Ni, Zr, Nb, Mo, Tc, Pd, La, Sm, Ta, Os, Ir, Pt, Th, Pa, U and their alloys can absorb many hydrogen atoms. The hydrogen states in the interstices of these metals are thought to be H^+, H, and H^-.

Two stable positions of the hydrogen atoms (H^+, H^-, or H) in the fcc (face centered cubic lattice), hcp (hexagonal close packed lattice), and bcc (cubic centered lattice) are pictured in Fig.5.19. One of them is called the "O-position" (octahedral intersticial position) and the other is the "T-position" (tetrahedral interstitial position). Well known examples are as follows.

fcc: Pd (a = 38.9 [nm]),

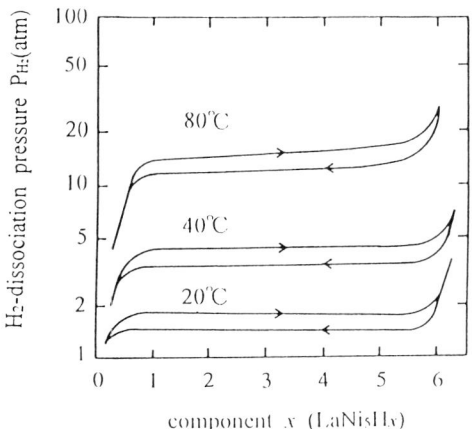

Fig. 5.20. PCT-curves for LaNi$_5$H$_x$.

Th (a = 50.7 [nm]),

hcp: Mg (a = 32.0 [nm], c = 52.0 [nm]),
Ti (a = 29.5 [nm], c = 47.3 [nm]),

bcc: Nb (a = 33.0 [nm]),
Ta (a = 33.0 [nm])

where a and c are the lattice constants indicated in Fig.5.19.

The general equation of hydrogen absorprion into a metal is

$$\frac{2}{y-x} MH_x + H_2 \rightleftarrows \frac{2}{y-x} MH_y + Q. \tag{5.14}$$

Putting x = 0 and y = 1, we have Eq.(5.13).

Discussing the characteristics of metal hydrides, three thermodynamic parameters are necessary. The first is the numerical ratio between hydrogen atoms and metallic atoms which is denoted by C (< 1) at constant temperature, and the second and third are the pressure (P) and the temperature (T), respectively. The pressure P_i pushing hydrogen into the metal and the pressure for dissolving hydrogen from the metal P_o are not equal to each other,

$$P_i > P_o.$$

Three parameters, P, C, and T, determine the characteristic of a metal hydride and this is called the "*PCT*- curve".

The characteristics of a representative metal hydride, LaNiH$_x$, is shown in Fig.5.20. The dissociation pressure is sensitive to temperature, and hysteresis phenomena are appreciable. The upper and lower pressure curves represent the charging and discharging cases, respectively.

The thermodynamic phase rule is maintained between the gaseous state (H$_2$) and solid state (M) in metal hydrides.

The degree of freedom (f) of the metal hydride, the number of components (c), and the number of phases (p) satisfy the phase rule :

$$f = c - p + 2, \tag{5.15}$$

in other words, f variables can be arbitrarily chosen among P, C, and T. In the present case, we have $f = 2$ because $c = 2$ and $p = 2$.

Next we note that the P vs x curves are flat in the figure, which indicate that hydrogen gas is effectively stored. Metal hydride systems having a flat pressure characteristic are classified into three categories :
  1. Titanium-iron alloys
  2. Lanthanum-nickel alloys
  3. Magnesium-nickel alloys.

Some representative metals and alloys are listed in Table 5.3. Besides them, more than one hundred metal hydrides have been manufactured to examine their characteristics.

(ii) *Hydrogen storage.* To store hydrogen in a metal hydride, a small value of Q is first required because cooling and warming need energy. In addition the flat length in the P vs C plot at room temperature should be long enough to store hydrogen.

Table 5.3. Characteristics of some metal hydrides

Metalhydride	H$_2$ wt%	Dissociation pressure [10^3 hPa]	Evolved heat
LiH	12.7	1(894℃)	−43.3
Mg$_2$NiH$_{4.0}$	3.6	1(250℃)	−15.4
LaNi$_5$H$_{6.0}$	1.4	4(50℃)	−7.2
MmNi$_5$H$_{6.3}$	1.4	34(50℃)	−6.3
MmNi$_{4.5}$Mn$_{0.5}$H$_{6.6}$	1.5	4(50℃)	−4.2
MmNi$_{4.5}$Cr$_{0.5}$H$_{6.3}$	1.4	14(50℃)	−6.1
MmNi$_{4.5}$Mn$_{0.5}$Zr$_{0.05}$H$_{7.0}$	1.6	4(50℃)	−7.9
TiFeH$_{1.9}$	1.8	10(50℃)	−5.5
TiCo$_{0.5}$Mn$_{0.5}$H$_{1.7}$	1.6	1(90℃)	−11.2
Ti$_{0.75}$Al$_{0.25}$H$_{1.5}$	3.4	1(100℃)	−11.3
VH$_2$	3.8	8(50℃)	−9.6
Ti$_{1.2}$Cr$_{1.2}$V$_{0.8}$H	3.0	4(150℃)	−9.1

Finally, the metal should be light enough to be equipped to vehicles driven by hydrogen fuel.

At present, alloys with mischmetal, Mm, are commonly utilized in hydrogen storage. The characteristics are shown in Table 5.3. Dissociation pressures and heat are compatible with ambient conditions. However, the price of Mm alloys is so high that heavier ferrous alloys are used instead. Metal hydride storage of hydrogen is safer than using liquefied or pressurized gas vessels, as shown in Table 5.4.

Hydrogen fueled cars have been developed since the begining of the 1970s. Some of them utilize liquefied hydrogen, while others use metal hydride to store hydrogen.

Using the evaporated cold H_2 gas to refrigerate its load, a refrigeration truck fueled by hydrogen from LH_2 was recently developed in Japan. However, the lack of LH_2 in countries where electrical power is costly limits the development of cars fueled by LH_2. Cars with metal hydride storage vessels have also been developed in industrial countries such as Germany, Japan, and the U.S.A. Such cars are expected to proliferate because of their non-pollutant exhaust gases.

(iii) *Metal hydride heat pump.* Most air-conditioners use flon gas as their heat medium. Flon has been feared to destroy the ozone layer in the stratosphere. Therefore, regulations regarding the utilization of flon are imposed in industrial countries. Alternative heat media or systems are urgently required.

Metal hydride air-conditioning and heating mechanisms are shown in Fig.5.21. Taking room heating as an example, we explain the system. Metal hydride M_1H is cooled by the room temperature and then has more capacity to absorb hydrogen from another metal hydride, M_2H. When absorbing hydrogen, M_1H's temperature rises, evolving the reaction heat that is emitted from M_1H into the room.

The refrigeration mechanism is similar. Ambient cold air cools M_2H, which then has more capacity to absorb hydrogen from another metalhydride M_1H. As the M_1H loses hydrogen, it absorbs heat from the room which is thus refrigerated.

Table 5.4. Storage media of hydrogen

Medium	Density	Hydrogen [wt%]	Number of atoms [10^{22}/c.c.]
LH_2	0.07	100	4.2
gas H_2 (1.5×10^5 h Pa 20℃)	0.012	100	0.38
$FeTiH_{1.74}$	6.1	1.52	5.5
$LaNi_5H_6$	8.25	1.37	6.76

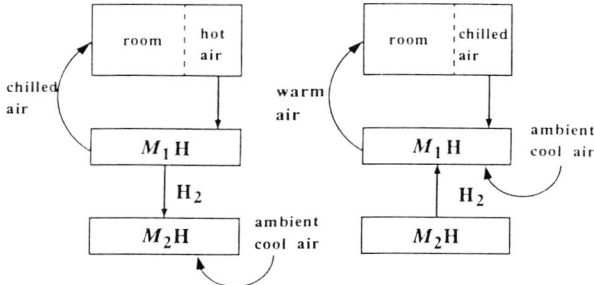

Fig. 5.21. Air conditioning using metalhydride cycle system.

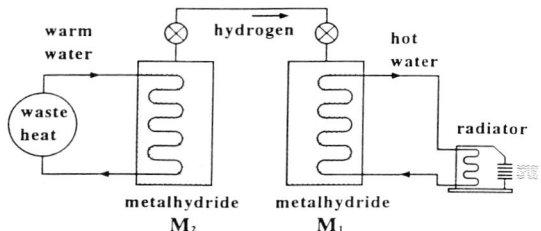

Fig. 5.22. Heating system using metal hydride.

A diagram showing the metal hydride heating system is given in Fig.5.22. One may expect that current refrigeration and heating systems will be replaced by metal hydride systems if more effective, inexpensive, and reliable alloys are manufactured in the near future.

(3) Superconductors

Superconductivity has been one of the most interesting topics of the twentieth century. Its application to energy systems has also been investigated but not been tried on a full scale. A brief survey of superconductivity and the essential points for its application are discussed in this section.

(i) *Brief survey.* The critical temperature (T_c) is defined as the temperature below which the electrical resistivity of a metal or an alloy becomes zero. This state is called the superconducting state and has three distinct characteristics. These are (1) electrical resistivity is zero, (2) perfect diamagnetism is realized, *i.e.*, magnetic flux density (B) is zero in the superconductor, (3) entropy is zero, so that no thermoelectric phenomenon occurs.

However, there exist appreciable effects on the characteristics of the superconductor in the presence of a magnetic field. Increasing the applied magnetic field, the superconducting state disappears and is replaced by the normal state.

At the critical magnetic field, H_c is given by

$$H_c(T) = H_0[1 - (T/T_c)^2], \tag{5.16}$$

where H_o is the critical temperature at $T = 0$.

The superconducting state excludes magnetic flux which penetrates to only a thin part of the outermost skin of the superconductor ($= 10^{-6}$ [cm]); this phenomenon is called "Meissner's effect" and the thickness (λ_L) is called the "penetration depth" given by

$$L^2 = m/(\mu_o n_s e^2), \tag{5.17}$$

where m, e, μ_o, and n_s are electronic mass, its charge, magnetic permeability of a vacuum, and the density of superconducting electrons, respectively.

If a superconducting straight wire is placed in a magnetic field in a parallel direction, along the wire axis, and shows perfect diamagnetism until the critical field is reached, the superconductor is called a "first kind superconductor" or "soft superconductor", because this type of material is soft (e.g. Pb, Sn, Hg, *etc.*). Most pure metals belong to this type. There exists another kind of superconductor called "second kind superconductors" or "hard superconductors" such as Nb and most alloys. These materials have two kinds of critical fields H_{c1} and H_{c2}. When the magnetic field increases above the lower critical field H_{c1}, magnetic flux lines penetrate into the wire and the wire becomes normal when H attains the upper critical field H_{c2}, that is, magnetic flux density n becomes

$$n = B/\phi_o, \quad B = \mu H_{c2}. \tag{5.18}$$

where B is the applied magnetic flux density and $\phi_o = h/(2e) = 2.07 \times 10^{-15}$ [Wb/m^2] is the unit of magnetic flux and is called "flux quantum", or "fluxoid". The fluxoid enters the wire as a line and is thus called a "flux line". The superconducting state is realized by the charge carrier, a so called "Cooper pair". A Cooper pair consists of two electrons that have anti-parallel wave vectors and anti-parallel spins, existing near the Fermi surface of the metal. The Fermi energy is defined as the chemical potential of a electron at $T = 0$ [K]. A Fermi surface is a surface formed by equi-Fermi energies. Electrical conduction in metals is due to electrons near the Fermi surface.

The state between H_{c1} and H_{c2} is called a "mixed state" where the number of flux lines in the superconductor varies, being proportional to the applied field. Unless the flux lines move, the resistivity of the mixed state is, in principle, zero. However, flux lines are sbject to the "Magnus force" in the direction perpendicular to both electric current and the magnetic field generated by the current itself. Therefore flux line pinning is contrived, for example.

(ii) *Applications.* To utilize superconductivity, higher critical temperatures (T_c) and higher upper critical fields (H_{c2}) are necessary. In addition, a large density of electric current flowing in a wire generates a strong magnetic field, and a current greater than a critical value will break the superconducting state. This critical value is called the "critical current" (I_c).

198   Energy Technology

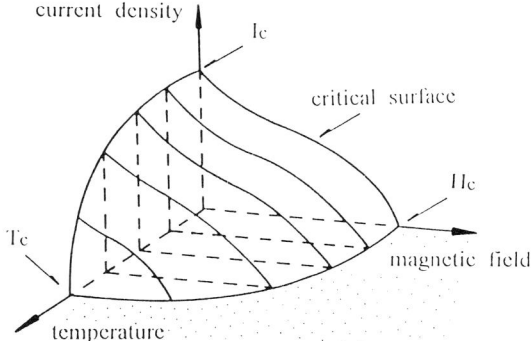

Fig. 5.23. *T-H-I* critical surface of superconducting state.

Three-dimensional plotting of the *T-H-I* for the superconducting state is shown in Fig.5.23. The interior of the *T-H-I* critical surface is the superconducting state, while the exterior is in the normal state. Higher T_c, H_c, and I_c are necessary for practical use.

The critical temperature is given by

$$T_c = 1.04\Theta \exp\{-1/[N(0)V]\}, \tag{5.19}$$

where Θ is the Debye temperature, $N(0)$ is the electron density of states at the Fermi surface, and V is the average coupling potential between electron and phonon.

The critical temperature is always lower than the Debye temperature and inceases with $N(0)$ and V. A strong interaction potential V is a property of the lattice structure of the metal, which gives the material a high T_c.

Most hard superconductors are difficult to manufacture into wire for use in energy equipment. Only the alloys, NbTi, Nb$_3$Sn and V$_3$Ga, are commonly utilized practically. Critical current density (I_c) vs upper critical field (H_{c2}) curves are shown in Fig.5.24. Typical values of T_c and H_{c2} for both Nb$_3$Sn and V$_3$Ga are about T_c = 15 - 23 [K] and H_{c2} = 20 - 40 T. Here the unit T is used. It is defined as 1 [T] = 1 [Wb·m^2].

A hydrogen bubble chamber used by CERN (European Organization for Nuclear Research) is equipped with a large magnet using a superconducting current. The material is a Nb-Ti alloy whose I_c-H_{c2} characteristic at T = 4.2 [K] (temperature of liquid helium) is plotted in Fig.5.24. It can be seen that V$_3$Ga and Nb$_3$Sn are electrically superior to NbTi.

The magnet, whose inner diameter is 4.7 [m] and height is 3.5 [m] generates a magnetic field as high as 3 [T] and a stored energy of 8 x 10^5 [kJ].

On the other hand, a magnet made of Nb$_3$Sn and V$_3$Ga has been utilized in a laboratory of the National Research Institute of Metals (Japan). This magnet

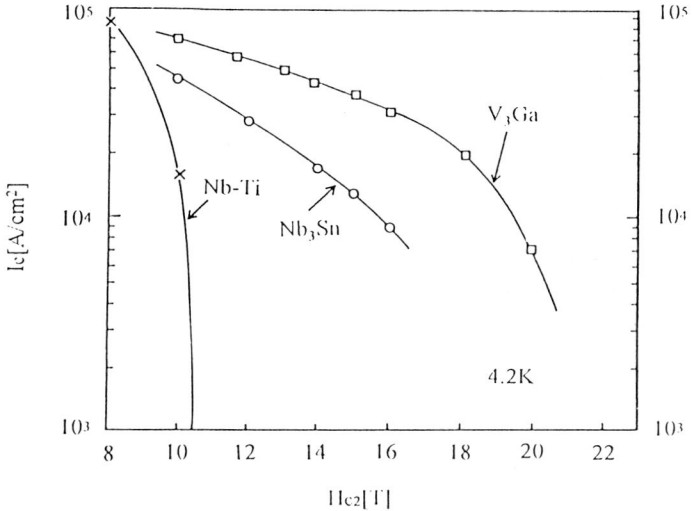

Fig. 5.24. I_c vs H curves for Nb$_3$Sn and V$_3$Ga.

generates a magnetic field of strength 17.5 [T] in the interior space of a 3 [m] magnetic coil. The value 17.5 [T] is the record of strength up to 1993.

(iii) *High-T_c superconductors.* J.G. Bednorz and K.A. Müller have discovered a ceramics superconductor La$_{5-x}$Ba$_x$Cu$_5$O$_{5(3-y)}$ with a critical temperature of as high as T_c = 35 [K] in 1986[6]. Since then many high-T_c superconductors of the same class have been found. For example, the elements La and Ba are replaced by Y, Nd, Sr, and Ce in Bednorz's sample of copper oxide. More than 150 kinds of high-T_c superconductors have been investigated prior to 1993.

The structure of superconductors containing CuO$_2$ are depicted in Fig.5.25 and classified as follows.
1. *Plane type.* Copper and oxygen atoms exist in the same plane (Fig.5.25, a).
2. *Pyramid type.* Oxygen atoms are situated at the top of a pyramid whose base plane is formed by one copper and four oxygen atoms (Fig.5.25, b).
3. *Octahedron type* (Fig. 5.25, c).

It is well known that CuO$_2$ is a ceramic insulator and no electric current flows in the plane formed by four oxygen atoms around one copper atom, but if surplus charge (positive or negative) is doped, then this structure give rise to superconductivity at high temperature.

The crystal structure of high-T_c superconductors, La$_{2-x}$Sr$_x$Cu$_4$ and Ln$_{2-x}$Ce$_x$CuO, are shown in Figs.5.26 (a) and (b), respectively. Comparing these two figures, we can see why these two materials are high-T_c superconductors. From this example, it must be noted that the electronic and lattice structures of the material are decisive for determining high-T_c superconductivity.

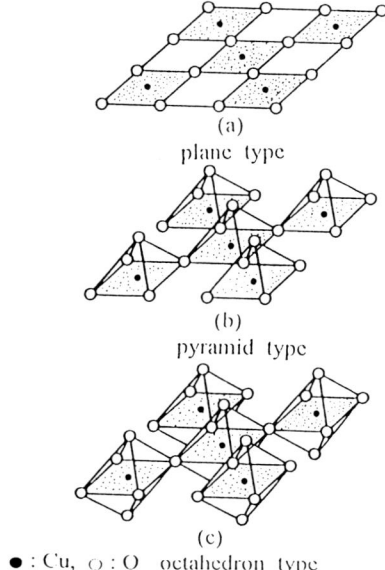

● : Cu, ○ : O   octahedron type

Fig. 5.25. Three types of two-dimensional CuO_2 plane.

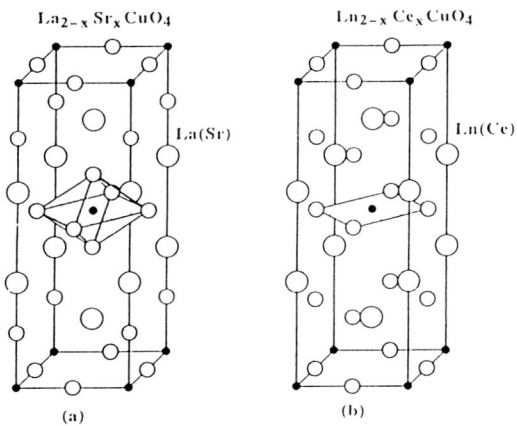

Fig. 5.26. Crystal structure of $La_{2-x}Sr_xCuO_4$(a) and $Ln_{2-x}Ce_xCuO_4$(b).

The main points of structural chemistry, micro-structure and electrodynamics of those superconductors are concisely described in A.M. Portis' text (*Electrodynamics of High-temperature Superconductors*, 1992, World Scientific Publishing Co. Ltd.).

The 20th meeting of the "LT-Physics" conference (LT 20) was held in Eugene (U.S.A.) in August of 1993. At this meeting, C.W. Chu reported that the highest critical temperature attains 146 [K] (- 27°C) for a system of superconductors, $HgBa_2Ca_{n-1}Cu_nO_y$ (n = 1 - 4).

The basic difficulty of applying these high-T_c superconductors to practical use is that no theoretical explanation has yet been possible and it is therefore almost impossible to reliably predict the behaviour of machines equipped with these superconductors.

5-4. Innovation of Systems

(1) Introduction

Not only energy systems but also every technological system has been innovated as a result of free competition based upon economy. Innovation nowadays is based on similar lines but the following restrictions apply so that free competition is not always a factor.

1. Alternative energy system to the petroleum system.
2. Constraints from global environmental problems.
3. Constraints from depletion of oil, gas and uranium.
4. Economical limitation by mitigation and adaptation to environmental problems.
5. Limitations for the security of an information society.
6. Conditions resulting from the life-styles of individual people.
7. Will Kondratiev's cycle still be valid in the future ?

Previously, alternative technologies came onstage, driven by economy, convenience (such as automation or labor saving), and fashion. It is said that "necessity is the mother of invention" and innovation up to now has followed this proverb. Nevertheless, technology innovation sometimes awaken the needs. Taking into account the situation described above, the constraints or limitations are closing to us. Some constraints resulting from global environmental problems are, as already mentioned, the atmospheric temperature rise that will raise the level of the sea surface, enlarge the desert area, and bring about unusual climatic conditions, and so on.

Alternative energy systems to the present oil and gas systems must be actively encouraged and developed. The hydrogen energy systems discussed in this text are leading candidates for such development.

The next constraint comes from the depletion of oil, gas, and uranium. Coal and renewable natural energies will be the next main energy sources. Petroleum is a compound of carbon (C) and hydrogen (H_2). If this is depleted, then coal and the hydrogen manufactured from water may play important roles as alternatives to oil.

Our society has moved into an **information age** with unexpected speed. Much information is flowing throughout society with the spread of valuable information being very fast, and shared equally all over the world. Although no projected value can be found, such spreading is, undoubtedly, a key to world peace. Nevertheless the information society is supported by an electrical power utility network that is produced by primary energy. World security depends upon the security of energies.

Individual life style will decide the energy demands from now on so as to encourage the active development of non-depletable and clean energy resources.

(2) Hydrogen energy systems
(a) Thermochemical water-splitting.[4, 36]
To split water, free energy (G) as well as heat (Q) is required. The total energy, that is, the enthalpy (H), changes by ΔH when water is split and is given by

$$\Delta H = \Delta G + \Delta Q. \quad \Delta Q = T\Delta S.$$

This equation has been presented a few times throughout the text and is valid for an isothermal reversible process.

If free energy is provided by electrical energy and photon energy, the water-splitting systems are called electrolysis and photolysis (photoelectrochemical splitting), respectively.

Here, a case where the free energy is derived from chemical energy is discussed. This process is called "thermochemical water-splitting".

Enthalpy *vs* temperature curves for water and its split state are given in Fig.5.27, where entropy *vs* temperature curves are also plotted. These curves are approximated by straight lines. The region of negative enthalpy is the state for water, H_2O, that is liquid or vapor, while the positive region shows the split state, $H_2 + \frac{1}{2}O_2$. An intermediate unstable state is shown by the broken line.

The crossing point between the negative enthalpy line and the temperature axis is denoted by T^*, which is 4,150 [K], then we have

$$T^* = \Delta H / \Delta S, \quad (5.20)$$

At this temperature no free energy is needed to split water. This water-splitting is called "direct thermal water-splitting".

At temperatures lower than T^*, free energy is required. For example, the ratio of free energy to heat at $T = 2,600$ [K] is $\Delta G/\Delta Q = 2/3$ as shown in Fig.5.27, where the values $\Delta Q = 2,600$ [K] x 11.6 [kcal/(mol·K)] = 30.16 [kcal/mol] and $\Delta G = 30.3$ [kcal/mol] are obtained.

For electrolysis at low temperature, the entropy change and temperature are so low that the quantity of electrical energy (ΔG) occupies more than 85 % of total energy (ΔH).

Now we notice a dotted cycle line starting from I which goes through path A-B-C-D-E-M-N-F. This cycle is the thermochemical water-splitting process. Let us take an example of the thermochemical water-splitting cycle as follows:

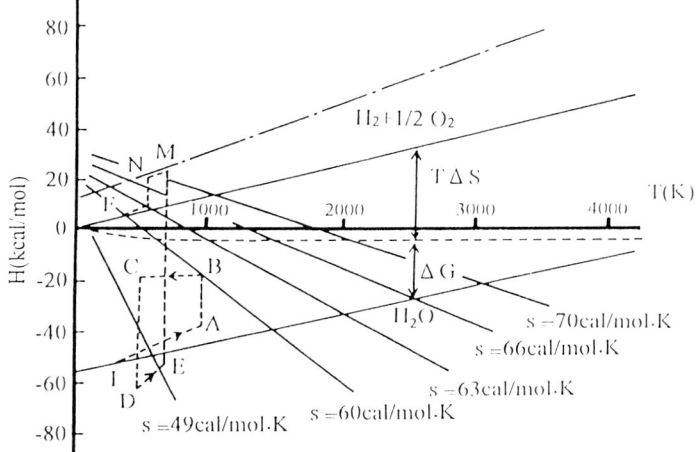

Fig. 5.27. Enthalpy *vs* temperature diagram for water-splitting.

$$\left.\begin{array}{c} X + Y + H_2O \rightarrow XH_2 + YO \\ XH_2 \rightarrow X + H_2 \\ YO \rightarrow Y + \frac{1}{2}O_2 \end{array}\right\} \quad (5.21)$$

where X and Y are the oxidization and reduction catalysts, respectively. Each reaction advances at temperatures as high as 1,000 [K].

From the starting state (I, 25°C, 10^3 [hPa]), the first reaction system corresponds to I-A-B. The separation of both products is shown by D. The second and the third reactions are denoted together by E-M, and the separation of H_2 and $1/2 O_2$ from the mixed system is shown by M-N-F. State M-N is an unstable state in which gasified hydrogen and oxygen are involved.

The sulphur-iodine cycle is a typical cycle and is shown by [T. Ohta, *Int. Jr. Hydrogen Energy* **13**(1988) 33],

$$\left.\begin{array}{c} SO_2 + 2H_2O + xI_2 \rightarrow H_2SO_4 + 2HI_x \\ H_2SO_4 \rightarrow H_2O + SO_2 + \frac{1}{2}O_2 \\ 2HI_x \rightarrow H_2 + xI_2 \end{array}\right\} \quad (5.22)$$

The thermal efficiency of the thermochemical cycle is unexpectedly high and is discussed in the following.

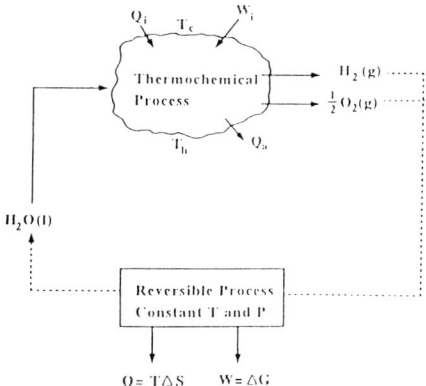

Fig. 5.28. Diagram of the thermochemical process.

In Fig.5.28, we show a diagram of the thermochemical process. A heat input Q_i at higher temperature T_h and work W_i are given to the system to split water. The unavailable heat Q_a is exhausted which has lower temperature T_c. Due to the first law of thermodynamics, we have

$$Q_i + W_i = Q_r + Q_a + W_r, \qquad (5.23)$$

where Q_r and W_r are heat and work needed in a reversible process. Therefore the relationships

and
$$\left. \begin{array}{l} Q_r = T_c \Delta S_o \\ W_r = \Delta G_o \\ \Delta H_o = Q_r + W_r \\ = \Delta G_o + T_c \Delta S_o \end{array} \right\} \qquad (5.24)$$

hold. In Eq.(5.24), the suffix o indicates the standard state (= 293 [K], and P = 10^3[hPa]).

Due to the second law of thermodynamics, we have

$$(Q_a + Q_r)/T_c - Q_i/T_h = 0. \qquad (5.25)$$

The overall thermal efficiency is defined by

$$\eta_o = \frac{\Delta H_o}{Q_i + W_i/\varepsilon} \qquad (5.26)$$

Table 5.5. Hydrogen energy systems outlook

Primary systems	Fossil fuels	Nuclear power	Solar energy	Biological	Hydro power	Wave energy	Waste heat	Wind energy	Gothermal	Ocean thermal gradient	Density gradient
conversion	cracking reforming		free energy + thermal energy thermochemical-splitting, electrolysis photolysis								
secondary system	utility electric power lines hydrogen energy systems										
tertiary system (utilization system)	**transport** **storage**										
	energy (combustion) ○ semiconductor refinery ○ iron & non-ferrous metal manufacturing ○ vehicles ○ jet, scram jet rocket ○ fuel cell ○ mixed city gas manufacturing	chemical feed stock (chemical reaction) ○ hydrogen protein ○ oil & fats industries ○ oil refining ○ methanol ○ ammonia synthesis	heat medium (reversible conversion) ○ metalhydride air-conditioning ○ metalhydride actuator (robot-system) ○ secondary battery									
safety & non-pollutants system											

where ε is the conversion efficiency from heat to work. Equations (5.23), (5.24), (5.25), and (5.26) give

$$\eta_o = \frac{(\Delta H_o \eta_c / \Delta G_o)}{1 + W_i(\eta_c - \varepsilon)/(\varepsilon \Delta G_o)} \quad (5.27)$$

with

$$\eta_c = (T_h - T_c)/T_h.$$

(b) Outlook for hydrogen energy systems. An outlook for hydrogen energy systems is shown in Table 5.5. One may anticitpate that hydrogen energy systems will replace the present oil system when oil is depleted or use is restricted due to global environmental problems.

Some technologies in Table 5.5 are already in effect while some of the others are under development or investigation. Hydrogen fueled jet planes[7] and a space plane equipped with a scram jet where the propellant is liquified hydrogen have been developed already.

(3) Refrigeration survey and alternatives

A metal hydride refrigeration system was introduced earlier although it is not yet in full scale practical use. The reason we first mentioned metal hydride refrigeration is, that manufacturing the present medium of flon is limited and will be forbidden in the near future in order to conserve the ozone layer in the stratosphere.

Flon molecules decompose **ozone molecules** in the stratosphere and ultra-violet solar radiation can then reach the earth surface inducing skin cancer to living creatures. This has been one of the more serious environmental problems.

Metal hydride refrigeration is not the only leading alternative and some more powerful candidates exist. Therefore in the following we shall mention every known method of refrigeration and entrust its evaluation to investigation.

Insofar as heating is concerned, it is believed that domestic needs will be met by a combination of solar beam collector, heat pump, and heat pipe. Electrical energy must be used optimally and not used for heating. Exergy loss is ats its highest when electrical energy is converted to heat (p.98).

Nevertheless demands for electrical power will increase so that its conservation will become vital. The motto of today's energy system is *"Heating from solar beam and refrigeration with minimum electricity"*.

(a) Thermodynamics of refrigeration.

(i) *Adiabatic expansion.* The most primitive refrigeration process is due to adiabatic expansion, where the *P-V* relationship is given by

$$pV^\gamma = \text{const.} \quad (2.2)$$

If an ideal gas in a state (V_1, P_1, T_1) expands adiabatically to another state (V_2, P_2, T_2), then we have the relationship between the two temperatures

$$T_2 = T_1(V_1/V_2)^{\gamma-1}$$

and the temperature difference is given by

$$\Delta T = (\gamma - 1)W/R, \tag{5.28}$$

where W is the work done by one mole of this gas against the external world.

Now, remember that refrigeration is the cycle undertaken in reverse order to Carnot's cycle, hence the Carnot efficiency is replaced by a coefficient of performance, that is defined by

$$c = Q_2/(Q_1 - Q_2)$$

$$= T_2/(T_1 - T_2) \quad (T_1 > T_2), \tag{5.29}$$

that is, the temperature drop is given by

$$\Delta T = T_2/c \tag{5.30}$$

From Eq. (5.28), an effective temperature drop is realized by (1) large expansion, (2) large value of ratio of specific heats, and (3) effective and large external work.

(ii) *Application of free energy.* In the case of adiabatic expansion, mechanical work is given to the medium gas to be compressed. Returning the energy to the external system, the system is refrigerated.

Now we consider the relationship between free energy and the temperature drop of a thermodynamic system.

The change of Gibbs' free energy for an isobaric reversible change for a refrigeration system is given by (p.138),

$$\Delta G = \Delta H - T\Delta S - S\Delta T$$

$$= \Delta U + p\Delta V - T\Delta S - S\Delta T. \tag{5.31}$$

Representing the reversible cycle by a suffix r, we have for the reversible process

$$\Delta U = W + Q$$

$$= W_r + Q_r$$

with
$$Q_r = T\Delta S$$

which is substituted into Eq.(5.31), to obtain

$$\Delta G = W_r + Q_r + p\Delta V - T\Delta S - S\Delta T$$

$$= W_r + p\Delta V - S\Delta T.$$

If this system is subjected to external work given by

$$W_r = -p\Delta V \tag{5.32}$$

then we have, for an isobaric reversible process,
$$\Delta T = -\Delta G/S \tag{5.33}$$

Equation (5.33) shows that the temperature drop is determined by the change of Gibbs' free energy. In other words, when applying the free energy ΔG, the magnitude of temperature drop is given by Eq.(5.33).

(iii) *Actual system.* Refrigeration systems need two kinds of energy, one is the energy contained in the heat medium, *e.g.*, chemical energy such as heat of vaporization, heat of sublimation, dilution heat, dissolution heat, and electronic energy as in Peltier cooling. Another is the energy needed to activate these chemical energies and transport them. Necessary energies for refrigeration as described above are listed in Table 5.6. We shall give brief comments on some important systems.

1. *Vapor-compression type refrigerator.* A schematic diagram showing a vapor-compression type refrigerator is shown in Fig. 5.29. A heat medium is compressed by the compresser and cooled by air or by circulated water, and the compressed gas then becomes liquid in the condensor. The liquid is vaporized by an expansion valve, absorbing heat from ambient space. After refrigerating the space, the medium is recovered and compressed again. This is one cycle of refrigeration and is shown in Fig.4.11 (the explanation is on p.138).

Conventional heat pumps have the same mechanism and a large coefficient of performance such as $c = 2.46$, because the conventional heat pump is of small scale and the refrigerated temperature drop ΔT is not large. The value of c corresponds to an energy efficiency of 70 %.

Most important is the **composition of the medium** which is usually ammonia, methylene chloride, and flons. Flons are chemical products where the hydrogen atoms, for example in methane (CH_4) or ethane (C_2H_6), are replaced by fluorine and chlorine atoms.The following are known: F-11[*] ($CFCl_3$), F-12 (CF_2Cl_2), F-21 ($CHFCl_2$), F-22 (CHF_2C), F-113[*] ($C_2F_3Cl_3$). CRC (Chlorofluorocarbon) with* are applied to large and massive refrigerators and methylene chloride is used by turbo-type refridgerators.

Table 5.6. Necessary energies for refrigeration

Refrigeration system	Medium's energy	Medium activating & driving energy (free energy)
heat-pump(*) vapor-compression system	vaporization-heat (sublimation-heat)	pressure (compression)
absorption-system	dilution-heat dissolution-heat	
adiabatic expansion	internal energy of gas	
adiabatic demagnetization(**)	magnetic interaction energy of spins	magnetic energy
Peltier cooling	electronic energy	D.C. current
laser cooling(**)	momentum of atom	laser-light
metalhydride	interaction-energy of metal with hydrogen	pressure or gradient

*Most popular refrigeration system.
**Topics of modern physics.

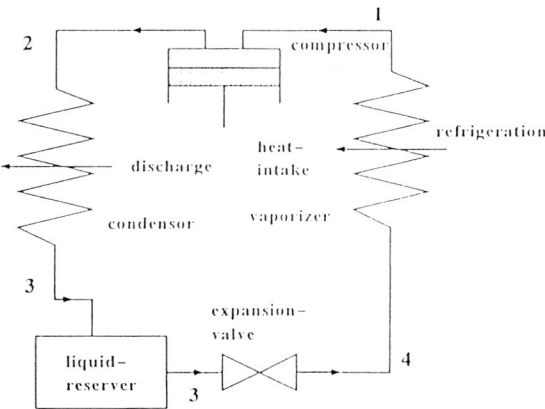

Fig. 5.29. System diagram of vapor-compression refrigeration.

As previously mentioned, developing refrigerators without flons, is a large task.

2. *Absorption (adsorption) type refrigerators.* The name differs according to the case. If the medium's vapor is directly absorbed into absorbent, then the name is absorption-type, while adsorption -type uses a solid adsorbant such as silica gel and Zeolite. A schematic diagram for an absorption type refrigerator is shown in Fig.5.30. A solution with a heat medium (solvent) and an absorbent (solution) is heated in a heater to separate the solvent vapor. The solvent is vaporized taking heat from the ambient space (refrigeration) and the vapor is then absorbed by the absorbent at the absorber. The absorber plays the role of reducing pressure.

Examples of combinations of solvent and solution are: water (solvent) with lithium (solution), ammonia (solvent) with water (solution), and sulphur dioxide (solvent) with silicate gel (adsorbent). Zeolite may be used as an effective adsorbent.

City gas fuel can be used to drive this type of refrigerator, however; this type has more functioning parts so that the body is more massive compared to the vapor-compression type.

3. *Adiabatic demagnetization.* Replacing the role of pV by $-MH$, where M and H are the magnetization and magnetic field, refrigeration by adiabatic expansion is possible with use of a proper magnetic substance.

Due to the analogous adiabatic ($S = 0$) relationship, we have

$$(\frac{\partial T}{\partial H})_S = -(\frac{T}{C_h})(\frac{\partial M}{\partial T})_H, \qquad (5.34)$$

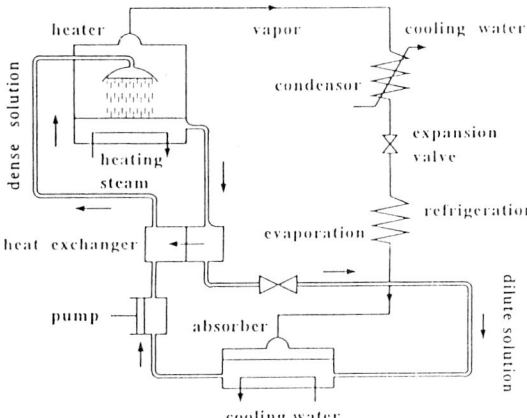

Fig. 5.30. System diagram of absorption refrigeration.

where C_h is heat capacity at constant magnetic field.

The partial differential $(\partial M/\partial T)_H$ is always negative for paramagnetic substances, so that $(\partial T/\partial H)_S > 0$, i.e., say magnetization produces a temperature rise, then adiabatic demagnetization generates a temperature drop (the reverse process). Physicists have tried to utilize this method since 1993 to attain a temperature below that of liquid helium (4.2 [K]).

Paramagnetic salts such as $Ce_2Mg_3(NO_3)_{12} \cdot 24H_2O$ are used in this type of refrigeration and the lowest temperature attained up to now is 2×10^{-3} [K] when refrigeration started from 1 [K] and the applied field is 1 - 6 [T].

The temperature 1 [K] is obtained by mixing ^3He with ^4He or by nuclear adiabatic demagnetization. The latter is realized by demagnitization of nuclear magnetism instead of atomic magnet.

The physics of this type refrigeration are as follows. An atomic magnetic material with magnetic moment μ in a magnetic field has its spin array put in order. If the magnet is removed, then the spin system absorbs the ambient heat to make a random arrangement. Rough estimates of the absorbing heat per mole of the magnetic material are $N\beta\mu H/R$, where β is a coupling constant between μ and H, and N is Avogadro's number.

4. *Electrically motive refrigeration.* Beside the Peltier refrigeration described already (p.69), some other refrigerating phenomena are known that have not yet been put to practical use. Applying a magnetic field adiabatically, strong enough to make a transition from a superconducting state to the normal state, refrigeration occurs as a result of superconducting electrons changing to normal electrons thus absorbing ambient heat.

Similarly, if we consider dipole molecules that are arranged in order (ferroelectric-materials) by a strong electrostatic field, and the field is then removed adiabatically, the system will be refrigerated.

5. *Ultimate low temperature.* When a particle of mass m is slowing to stopping with a very slow speed δv, its de Groglie wave length is decided by the uncertainty principle and is given by

$$\lambda = h/(m\delta v). \qquad (5.35)$$

A physical science group at Stanford University (U.S.A.)[25] has published the results of an experiment where a laser beam was directed at sodium atoms in a vacuum tube and their speed slowed down to 0.27 [mm/s]. This speed is so slow that the momentum is 71.97×10^{-31} [kg·m/s] and the corresponding de Broglie wavelength becomes 7×10^{-6} [cm], in the same realm as infrared light. Therefore it is possible to create interference between two similar de Broglie waves manufactured by laser beam control. The interfered beams achieve zero net amplitude and at such an interference, the Na atoms can hardly move. This state corresponds to temperatures as low as 24×10^{-12} [K].

(4) Frontiers and their heyday

Kondratiev[26] noted an interesting trend concerning the rise and fall of new technologies. A new technology gains more commercial acceptance and reaches a maximum if the technology succeeds in overcoming all difficulties and is welcomed by society. Then as it is gradually replaced by newly developed technology its amount of commercial trade tends to zero about 45 - 55 years after it was first introduced. This cycle is called "Kondratiev's cycle" and is shown in Fig.5.31. It is often cited when discussing innovation and frontiers of technologies. We shall make some comments on this cycle introducing concrete examples[31,39].

(i) *The eighteenth century.* The eigthteenth century was the century of the Industrial Revolution. The traditional relationship between technology and civilization was established during this century. The term "traditional" means that the technologies that survived won the day without any restriction or constraint. Research and development were performed in every effort to invent cheaper and more convenient technologies and discover more abundant and economical resources. Such traditional principles as described above are believed to yield to Kondratiev's cycle.

However, it can be said that free competition cannot exist in the future as mentioned previously. Thus we can hardly predict the tendency of the cycle after 2,000.

The most fundamental driving force for the **Industrial Revolution** was the steam engine invented by J. Watt (1736 - 1819). The invention of a condensor, which separates the steam (working substance) from cooling water, contributed to decisive increases in the efficiency of steam engines (1769).

The energy resource in eighteenth centrury was firewood which occupied more than 95 % in 1850. With the spreading of the steam engine, demands on iron rapidly increased and firewood, from which charcoal is manufactured, would be exhausted. Abraham Darby tried to utilize coke as an alternative to charcoal and succeeded in 1709. Since then coal became the main energy resource of industries, because charcoal can be manufactured from coal.

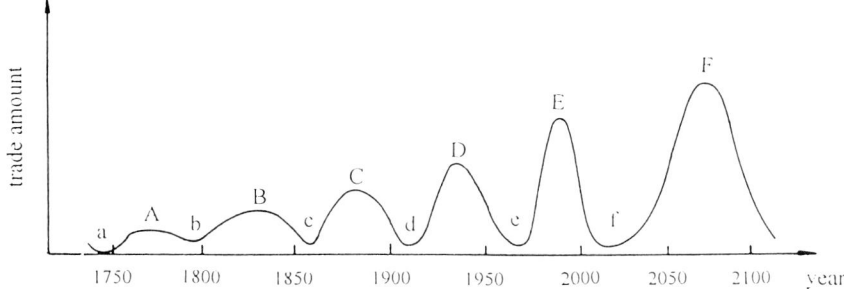

Fig. 5.31. Frontiers and their cycle (after Kondratiev[26]) a and A represent a valley and mountain where technology innovations are reared up and succeeded to attain maximum trade, respectively. The interval between consecutive peaks is about half a century.

(ii) *The nineteenth century.* Watt's engine had spread rapidly all over the world and small-sized engines were required by many kinds of industries. In 1887, Otto's four-stroke engine was developed, it was fueled by gas and was so economical and much quieter that many inventors also tried to invent small-sized engines. Meanwhile, G. Daimler and C. Benz, independently, had invented engines with greater speeds that were fueled by liquid fuel in 1883.

These events marked the start of the petroleum age, while no petroleum was spent on the world statistics in those ages yet.

G.S. Ohm (1784 -1854) discovered "Ohm's law" in 1826 and shortly afterwards J.R. Joule (1818 - 1889) found a numerical value for the mechanical equivalence of heat, in 1841, which was succeeded by J.R. von Meyer (1814 - 1879), who established the first thermodynamical law defining energy conservation in 1842. This period was full of interest and investigations into energy science and technology.

S.F.B. Morse (1791 -1872) invented the telegraphic instrument which was the pioneering technology of the information age. Communications between England and Europe started in 1851 with submarine cables. On the other hand, J.C. Maxwell published his theory of electromagnetic waves in 1855, and 33 years later H.R. Hertz (1857 - 1894) carried out experiments demonstrating wireless telecommunication by electric waves.

The applications of electricity are thus classified into two groups. One is energy utilization and another is communication. Of the two, energy application advanced rapidly. Examples include the invention of the battery in 1868 by G.Leclanche (1839 - 1882), establishment of the power generating station in 1881 by T.Edison (1847 -1931), the A.C. generating station by Westinghouse Co. Ltd. in 1886, and three phase A.C. current with 10 kW capacity transmitted in 1887.

Industry utilized a large amount of electrical energy in electric lamps, electric motors, telephones, and aluminum manufacturing.

Beside these energy related industries, photographic technology, ammonia synthesis and cement manufacturing were also developed.

(iii) *The twentieth century.* The frontiers of technologies in the twentieth century can be classified into three categories. (1) energy, (2) material, and (3) information. The most remarkable demand on energy at the begining of the twentieth century was the rapid increase in utilization of gasoline and oil due to the widespread use of vehicles. For example, the first jetplane flew only 34 years after the Wright brothers invented the first airplane. The rise of airplane transportation is one of the most distinctive developments of twentieth century civilization. This was due to the invention of light and strong materials like duralumin (1907) and heat-proof alloys such as Inconel 700 and 713C.

More than 70 % of the world's primary energy of world demands were met by coal in 1925, while petroleum and gas occupied more than 50 % of world needs in 1950. It can be said that the twentieth century was the century of coal → petroleum and iron → plastics.

A few new materials that arrived in the twentieth century include Bakelite (invented in 1907), Acetate (1913), Polyethylene (1934), Nylon (1938), and other complex materials. All of them are manufactured from fossil fuels, but recently several ceramics have been recovered from the stone age to play important roles in frontier technologies. Glass-fibre is also indispensable in laser communication technology. These ceramics cannot be manufactured unless enough electrical energy is provided.

In 1942, the first nuclear reactor was successfully operated in E. Fermi's laboratories in Chicago. Since then nuclear energy has increasingly contributed to the generation of electric power. At the end of June 1991, there were 422 nuclear power stations in the world, and their generated electrical power totalled 343.41×10^6 [kW], which accounts for 16,470 of the world's generated power.

Since the invention of highly integrated circuit devices using CI, VLSI, and mega class DRAM microelectronics as well as computer technologies have advanced rapidly.

One of the most feasible developments for frontier technology in the twenty-first century which requires little energy and materials will be life technologies such as genetic engineering. We expect that the trend of technology development and the supporting energies will be directed towards environmental protection and resource management, placing the preservation of life at their center. The frontiers of development will be those described in this text.

Appendices

UHV power cable with 1MV,
the world highest voltage

A-1. Nomenclature and Units

A :	ampere	[A]		F :	Faraday's constant	[C/mol]
	amplitude				force	[N]
	cross section area	[m^2]			generalized force	
	technology development factor				Helmholtz's free energy	[J]
a :	activity			f :	force	[N]
	length	[m]		f:	degrees of freedom	
	quality factor					
	reducing rate			G :	Gibbs' free energy	[J]
B :	barrel (1B = 0.159 kl)			g :	gravitational acceleration	[m/s^2]
C :	capacitance	[F]		H :	enthalpy	[J]
	Coulomb	[C]			height	[m]
	electric capacity	[F]			magnetic field	[A/m]
	heat capacity	[J/(kg·K)]		h :	heat of dissolution	
c :	coefficient of performance				Planck's constant	[J·s]
	flow of consumption per capita				hour	
	light velocity	[m/s]		I :	electric current (density)	[A] ([A/m^2])
D :	day				construction cost per kW	[$/kW]
	electric displacement	[C/m^2]				
	development cost	[$/(kW·h)]		J :	Joule	
d :	distance	[m]				
	annual decline rate	[kg/Y]		j :	current density	[kg/(m^2·s)] [A/m^2]
E :	electric field	[V/m]				
	exergy	[J]		K :	capital	
	work	[J]			elastic constant	
e :	efficiency (vector)			k :	Boltzmann's constant	[J/k]
	electronic charge	[C]			coupling constant	[J/m^2]

$L:$ diffusion length [m]
electric induction
[H = V·s/A = Wb/A]
inductance [H]
penetration depth [m]
preindustrial time (CO_2-emission)

$l:$ length

$M:$ mass [kg]
molar density [1/mol]
quantity

$m:$ mass [kg]
minute

$N:$ Avogadro's number [1/mol]
number

$n:$ particle number density (electron, positive hole *etc.*) [$1/m^3$]

$P:$ energy quantity
population
price of primary energy per l[t]
pressure [Pa]

$p:$ power density [$W/m^3, W/kg$]

$Q:$ electrical charge [C]
flow quantity
heat [J, kcal]
quality value

$q:$ electrical charge
annual output quantity [1/Y]

$R:$ electrical resistance [W]
investment
radius [m]
universal gas constant [J/(K·mol)]
searching cost [$/(kW·h)]

$r:$ displacement vector

$S:$ cross section area [m^2]
entropy [J/K]
Seebeck's coefficient (thermo-electric power) [V/K]
subsystem
entropy production rate
[J/(K·mol)]

$s:$ second [s]

$T:$ absolute temperature [K]
time [s, m, D, Y]

$T:$ torque [N·m]
transfer matrix

$t:$ time [s]

$U:$ internal energy [J]

$u:$ magnetic moment [Wb·m]

$V:$ electric potential [V]
volume [m^3]
average Coulomb potential [V]

$v:$ velocity [m/s]

$W:$ energy [J]
wattage [J/s],[W]
work [J]
work function [J, eV]
operating cost [$/(kW·h)]

X :	impact due to climatic change component along x-direction		μ :	chemical potential [J/mol] emission control rate magnetic permeability [H/m] magnetic moment [m/(s·V)] mobility(electron, positive hole) [m^2/(s·V)]
x :	position			
Y :	component along y-direction year [Y]			
Z :	component along z-direction figure of merit [1/K]		ν :	frequency [1/s]
			ξ :	extensive variable
α :	coefficient phase angle [rad]		ξ :	generalized displacement
			Π :	Peltier's coefficient [J/A]
β :	damping coefficient elasticity of output to capital		ρ :	density [kg/m^3] electrical resistivity [W·m]
Γ :	ranking of transportation		σ :	electrical conductivity [1/(W·m), s/m] Stefan-Boltzmann's constant [J/K^4]
γ :	ratio of specific heats (= C_p/C_v) elasticity of output			
ϵ :	dielectric constant [C/mv] solar constant [W/m^2]		τ :	collision time [s] confinement time [s] half life time [Y] life time [s, Y]
ζ :	phase angle [rad]			
η :	efficiency [%] intensive variable viscosity [poise, stokes, m^2/s, Pas]		ϕ :	work function [J]
			Ω :	ohm [V/A] thermodynamic probability
θ :	angle [rad]		w :	angular velocity [rad/s]
κ :	heat conductivity [W/(K·m)]		W_c :	cyclotron frequency [rad/s]
λ :	energy transfer rate wavelength [m, nm, A]			

A-2. Physical Constants

standard gravitational constant	g	9.806	m/s^2
speed of light in a vacuum	c	2.99792458	10^8 m/s
permeability of free space	μ_0	1.25663706	10^{-7} H/m
permittivity of free space	ϵ_0	8.85418782(5)	10^{-12} F/m
elementary charge	e	1.6021892(46)	10^{-19} C
	e^*	4.803242(14)	10^{-10} e.s.u.
Planck's constant	h	6.626176(36)	10^{-34} J·s
$h/2\pi$	$\hbar$	1.0545887(57)	10^{-34} J·s
electron mass	m_e	9.109534(47)	10^{-31} kg
		5.4858026(21)	10^{-4} u
proton mass	m_p	1.6726485(86)	10^{-27} kg
		1.007276470(11)	u
neutron mass	m_n	1.6749543(86)	10^{-27} kg
		1.008665012(37)	u
atomic mass unit	amu	1.6605655(86)	10^{-27} kg
fine structure constant	α	7.2973506(60)	10^{-3}
Rydberg constant	$R\infty$	1.097373177(83)	10^7/m
Bohr radius	a_0	5.2917706(44)	10^{-11} m
Bohr magneton	μ_B	9.274078(36)	10^{-24} J/T
nuclear magneton	μ_N	5.050824(20)	10^{-27} J/T
magnetic moment of electron	μ_e	9.284832(36)	10^{-24} J/T
magnetic moment of proton	μ_p	1.4106171(55)	10^{-26} J/T
Avogadro's number	N	6.022045(31)	10^{23}/mol
universal gas constant	R	8.31441(26)	J/(mol·K)
Boltzmann's constant	k	1.380662(44)	10^{-23} J/K
Faraday's constant	F	9.648456(27)	10^4 C/mol
Stefan-Boltzmann's constant	σ	5.670	10^{-8} W/(m^2·K^4)
magnetic flux quantum	$\Phi_0 = h/2e$		
		2.07	10^{-15} Wb
absolute temperature at 0°C	T_0	273.15	K
standard atmospheric pressure	P_0	1.01325	10^5 Pa

u : amu

A-3. SI Base and SI Derived Units

Physical Quantity	Name of Unit	Symbol
length	meter	m
mass	kilogram	kg
time	second	s
electric current	ampere	A
thermodynamic temperature	kelvin	K
amount of substance	mole	mol

Physical Quantity	Name of Unit	Symbol	SI Unit
frequency	hertz	Hz	s^{-1}
energy	joule	J	$kg \cdot m^2/s^2$
force	newton	N	$kg \cdot m/s^2$
pressure	pascal	Pa	$kg/(m \cdot s^2)$
power	watt	W	$kg \cdot m^2/s^3$
electric charge	coulomb	C	$A \cdot s$
electric potential	volt	V	$kg \cdot m^2/(A \cdot s^3)$
electric resistance	ohm	W	$kg \cdot m^2/(A^2 \cdot s^3)$
capacitance	farad	F	$A^2 \cdot s^4/(kg \cdot m^2)$
inductance	henry	H	$kg \cdot m^2/(A^2 \cdot s^2)$
magnetic flux	weber	Wb	$kg \cdot m^2/(A \cdot s^2)$
magnetic flux density	tesla	T	$kg/(A \cdot s^2)$

Energy Related Conventional Units

Physical Quantity	Name of Unit	Symbol	Conversion
volume	barrel	B	0.159[kl]
mass	carbon equivalent ton	cet	$(11/3) \times (CO_2$ mass$)$
energy	oil* equivalent ton	oet	51.0×10^6 [kJ]
			12.2×10^6 [kcal]
	coal** equivalent ton	cet	38.6×10^6 [kJ]
			9.24×10^6 [kcal]
	oil equivalent liter	oel	9.8×10^3 [kcal]
	oil equivalent mega ton	oeMt	10^6 times cet
	kilo watt hour	[kW·h]	860 [kcal]
	equivalent primary energy		2,530 [kcal]

* Burner oil equivalent, ** Anthracite

A-4. Conversion of Energy Units (A)

	eV	K	cm^{-1}	$Hz[s^{-1}]$	erg	cal	J/mol	kcal/mol
1 eV	1	1.16049×10^4	0.80657×10^4	2.4180×10^{14}	1.06210×10^{-12}	3.8291×10^{-20}	9.64869×10^4	23.05263
1 K	0.86171×10^{-4}	1	0.69503	2.0836×10^{10}	1.38055×10^{-16}	3.2995×10^{-24}	8.31435	1.98645×10^{-3}
1 cm^{-1}	1.23981×10^{-4}	1.43879	1	2.9979×10^{10}	1.98630×10^{-16}	4.74739×10^{-24}	11.96258	2.85812×10^{-3}
1 $Hz[s^{-1}]$	4.1356×10^{-15}	4.7993×10^{-11}	3.3356×10^{-11}	1	0.66256×10^{-26}	1.58359×10^{-34}	3.99026×10^{-10}	9.53362×10^{-14}
1 erg	6.24181×10^{11}	7.24349×10^{15}	5.03449×10^{15}	1.50930×10^{26}	1	2.3901×10^{-8}	6.0225×10^{16}	1.43891×10^{13}
1 cal	2.61157×10^{19}	3.0307×10^{23}	2.10647×10^{23}	6.314978×10^{33}	4.1840×10^7	1	2.5198×10^{-4}	6.02252×10^{20}
1 J/mol	1.03641×10^{-5}	0.12027	0.083594	2.5061×10^9	1.66043×10^{-17}	3.96853×10^{-25}	1	2.3901×10^{-4}
1 kcal/mol	0.043379	503.41	349.88	1.04892×10^{13}	0.69497×10^{-13}	1.6604×10^{-21}	4.1840×10^3	1

(1) $1[J] = 10^7[erg]$, (2) $[cm^{-1}]$(unit of wave number) and $[Hz]$(unit of frequency) represent the photon energy;
(3) [J/mol] or [kcal/mol] is the unit of chemical potential, (4) [K](unit of temperature) represents the thermal energy.

A-5. Conversion of Energy Units (B).

Units	g mass (energy equiv.)	J	int J	cal	cal_{IT}	BTU_{IT}	kWh	HPh	ft-lb(wt)	ft^3-lb (wt)in.$^{-2}$	$l\cdot atm$
1g mass (energy equiv.)	=1	8.987552×10^{13}	8.98069×10^{13}	2.148076×10^{13}	2.146640×10^{13}	8.518555×10^{10}	2.496542×10^{7}	3.347918×10^{7}	6.628873×10^{13}	4.6033388×10^{11}	8.870024×10^{11}
1J	$=1.112650 \times 10^{-14}$	1	0.999835	0.2390057	0.2388459	9.478172×10^{-4}	$2.77777... \times 10^{-7}$	3.725062	0.7375622	5.121960×10^{-3}	9.869233×10^{-3}
1 int J	$=1.112834 \times 10^{-14}$	1.000165	1	0.2390452	0.2388453	9.479735×10^{-4}	2.778236×10^{-7}	3.725676×10^{-7}	0.7376839	5.122805×10^{-3}	9.870862×10^{-3}
1 cal	$=4.655328 \times 10^{-14}$	4.184*	4.183310	1	0.9993312	3.965667×10^{-3}	$1.6222... \times 10^{-6}$	1.558562×10^{-6}	3.085960	2.143028×10^{-2}	0.04129287
1 cal_{IT}	$=4.658443 \times 10^{-14}$	4.1868*	4.186109	1.000669	1	3.968321×10^{-3}	1.163000×10^{-6}	1.559609×10^{-6}	3.088025	2.144462×10^{-2}	0.04132050
1 BTU_{IT}	$=1.173908 \times 10^{-11}$	1055.056	1054.882	252.1644	254.9985	1	2.930711×10^{-4}	3.930148×10^{-4}	778.1693	5.403953	10.41259
1 kWh	$=4.005540 \times 10^{-8}$	3600000	3599406	860420.7	859845.2	3412.142	1	1.341022	2655224	18439.06	35529.24
1HPh	$=2.986931 \times 10^{-8}$	2684519	2684077	641615.6	641186.5	2544.33	0.7456998	1	1980000	13750	2694.15
1 ft-lb (wt)	$=1.508551 \times 10^{-14}$	1.355818	1.355594	0.3240483	0.3238315	1.285067×10^{-3}	3.766161×10^{-7}	$5.050505... \times 10^{-7}$	1	$6.9444... \times 10^{-3}$	0.01338088
1 ft^3-lb (wt)in.$^{-2}$	$=2.172313 \times 10^{-12}$	195.2378	195.2056	46.66295	46.63174	0.1850497	5.423272×10^{-5}	7.272727×10^{-5}	144	1	1.926847
1 $l\cdot atm$	$=1.127393 \times 10^{-12}$	101.3250	101.3083	24.21726	24.20106	0.09603757	2.814583×10^{-5}	3.774419×10^{-5}	74.73349	0.5189825	1

int. IT : international unit

A-6. Energy Consumption of Some Manufacturing Industries[40].

industry	end-use energy	
	thermal TOE/t	electrical 10^3 kWh/t
brewing	0.05	0.10
cement	0.08	0.11
dairy industy	0.15	0.5
milk processing	0.02	0.10
non-ferrous metal	0.07	0.30
paper and pulp	0.15	0.43
plastics	0.05	0.55
rubber	0.10	5.00
textiles (dyeing)	0.75	0.75
textiles (spining)	0.45	7.5

1) TOE: ton oil equivalent. Burner oil: 42,680 kJ/kg
2) 1 kWh = 860 kcal = 3,600 kJ

References

The contents of this book originate from my original books (published in Japanese). These are as follows:
- (1) Ohta, T. ; *Energy Systems* (1976, NHK Books)
- (2) Takahashi, H. and Ohta T. ; *Fundamental Theory of Energy* (1984, Ohm Pub. Co.)
- (3) Ohta, T. ; Energy and Hightechnology, Chap.1. in *Frontier of Clean Energy,* Ed. by H.Hironaka (1991,CEIO)
- (4) Ohta, T. (Ed.); *Solar-Hydrogen Energy Systems* (1979, Pergamon Press)
- (5) Ohta, T. ; *Hydrogen Energy* (1987, Morikita Pub. Co.)

References listed below are classified into two categories. The first group supplement the contents of this book and the second group are works that have appreciably contributed to the author's learning. The number of references is limited to fifty.
- (6) Bednorz, J.G. and Müller, K.A. ; *Z. Phys.* B**64**(1986)189
- (7) Brewer, G.D.; *Hydrogen Aircraft Technology* (1991, CRC Press Inc.)
- (8) Clingman, W.H. and Moore, R.G. ; *J. Appl. Phys.* **32**(1961)675
- (9) Close, F. ; *Too Hot to Handle* (1991, Princeton University Press)
- (10) Cole, T. ; *Science* **221**(1983)915
- (11) Culp Jr., A.W. ; *Principles of Energy Conversion* (1979,McGraw-Hill Book Co.)
- (12) De Beghi, G. and Dejace, J. ; *THEME Conference Proceedings,* S2-11 1974, University of Miami)
- (13) De Groot, S.R. ; *Thermodynamics of Irreversible Processes* (1952, North-Holland Publishing Co.)
- (14) *Electronics* (1979, Jun. 7)55
- (15) *Engineering News Records* (1972, Feb.)17
- (16) Fraas, A.P. ; *Engineering Evaluation of Energy Systems* (1982, McGraw-Hill Book Co.)
- (17) Fujishima, A. et al. ; *J. Chem. Soc. Japan* **72**(1969)108
- (18) Gold, T. ; *Power from the Earth* (1987, George Weidenfeld and Nicolson Ltd.)
- (19) Hammond, A.L., Metz, W.D., and Maugh II, T.H. ; *Energy and the Future* (1973, Am. Assoc. Adv. Science)
- (20) *Handbook of Chemistry and Physics* 72nd Ed. (1991-1992, CRC Press Inc.)
- (21) Heidt, L.J.and Mcmillan, A.M. ; *J. Chem. Soc.* **76**(1954)2135
- (22) Jensen, J. and Sorensen, B. ; *Fundamentals of Energy Storage* (1984, John Wiley & Sons)
- (23) Justi, E.W. ; *A Solar-Hydrogen Energy System* (1987, Plenum Press)
- (24) Kameyama, H. et al. ; *Analysis of Energy Flows in Chemical Process* (1978, University of Tokyo, in Japanese)
- (25) Kasevich, M. et al. ; *Phys. Rev. Letters* **66**(1991)2297
- (26) Kondratiev ; VDI-*Nachrichten* (1986,NO.9,2)
- (27) King, C.J. ; *Separation Processes* (1971, McGraw-Hill Book Co.)
- (28) Lawson, L.J. ; *Proc. IECEC* (1971)
- (29) Mallove, E.F. ; *Fire from Ice* (1991, John Wiley & Sons, Inc.)
- (30) Moss, T.S. ; *Optical Properties of Semiconductors* (1960, Butterworths Science Pubs.)
- (31) Mussan, A.E. and Robinson, E. ; *Science and Technology in the*

Industrial Revolution (1981, Gorden and Breach)
(32) National Academy of Science, National Academy of Engineering, Institute of Medicine ; *Policy Implications of Greenhouse Warning* (1991, National Academy Press)
(33) NATO Science Committee ; *Technology of Efficient Energy Utilization* (1974, Pergamon Press)
(34) News in *Physics Today* **17**(1973)3
(35) Nordhaus, W.D. ; *Science* **258**(1992)1315
(36) Ohta, T. et al. ; *Int. Jr. Hydrogen Energy* **10**(1985)571
(37) Ohta, T. and Homma, H. (Ed.) ; *New Energy Systems and Conversions* (1993, Universal Academy Press, Inc.)
(38) Oshida, I. ; *Lectures on Exergy* (1986, Solar Energy Research Institute, in Japanese)
(39) Pacy, A. ; *Technology in World Civilization* (1990, Basil Blackwell)
(40) Petrecca, G. ; *Industrial Energy Management: Principles and Applications* (1992, Kluwer Academic Pub.)
(41) Posa, J.G. ; *Electronics Review* (1980, Nov.)39
(42) Rant, Z. ; *BWK* **12**(1980)297
(43) *Science* **241** (1988)900
(44) Shibata, T. et al. ; *J. Phys. Soc. Japan* **14**(1957)227
(45) Spitzer, L. ; *Physics of Fully Ionized Gases* (1956, Interscience)
(46) Takahashi, H. ; *Ohyo Butsuri* **47**(1971)7,80 (in Japanese)
(47) Tanaka, K. et al. ; *Thermal Designing on Wick Return AMTEC Cells in New Energy Systems and Conversions* (Ed.) Ohta,T. and Homma, T. (1993, Universal Academy Press, Inc.)
(48) Tuck, C.D.S. ; *Modern Battery Technology* (1991, Ellis Horwood Ltd.)
(49) Wilson, A.H. ; *Theory of Metals* (1953, Cambridge University Press)
(50) Van der Aren, P.C. and Chelton, D.B. ; *The Liquefaction of Hydrogen in Technology and Use of Liquid Hydrogen* (Ed.) Scott, R.B., Denton, H., and Nicols, C.M. (1964, Pergamon Press)

Indices

I. Scientific and Technical Terms

The term itself does not always appear on the page
but indicates the basic content of that page

A
AMTEC, 188
Apollo project, 118,146
Atkinson's cycle, 61
absorption type refrigeration, 210
activity, 99
adaptation, 115
adenosine di(tri)phosphate, 14
adiabatic demagnetization, 210
--- expansion, 206
air density, 102
--- pollution, 18,107
ammonia, 208
anion, 150,183
anthracite, 16
available energy, 58,67

B
Bernoulli's theorem, 13,39
Betz's formula, 13,40
Brayton's cycle, 61
BWR(boiling water reactor), 24
batch system, 95,131,160
beta (b" -) alumina, 188
--- decay (b-decay), 23
bituminous, 16
breeder reactor, 25
built-in field, 84,178
burner oil, 19,223
butane, 138,Table1.3,Table3.1

C
Carnot's efficiency, 52,67,144,207
--- cycle, 50,207
CFC(chlorofluorocarbon), 32,42,210
COM, 128

Cooper's pair, 197
Curie Temperature, 55
Curie-Weiss constant, 55
CWM, 128
calcium carbonate, 110
calorique, 10
capacitance, 167
carbon dioxide, 19,35,98,103,104,
108,170,182
--- monoxide, 149
catalyst (catalyzer), 151,185,187
cation, 150,183
ceramics, 149,214
cet(carbon equivalent ton), 108,220
--- (coal ---), 220
chemical affinity, 8
--- energy, 7,140,150,155
--- potential, 9
--- wattage, 15,130
city gas, 118,151
clean energy, 130
coal, 16
--- combustion, 16
coefficient of performance, 207
cogeneration, 118,126
cohesive energy, 8,100
coil, 73,151,157
cold fusion, 27
combustion mechanism, 8,144
compression gas system, 136
condenser, 151
confirmed reserve, 18,26
conversion heat, 187
converter, 73,168
copper ferrocyanide, 96
cost, 104
--- of primary energy, 104

cost of secondary energy, 105
coupling constant, 59
covalent binding, 8
critical field, 196,197
--- current, 197
--- temperature, 196,198
cryogenic storage, 138
cyclotron frequency, 77

D
De Broglie wave, 212
Debye temperature, 198
Dember's effect, 81
DICE model, 112
Diesel's cycle, 61
--- generation, 126
Doulong's formula, 16
dam system, 35
delta method (D-method), 166
density energy, 9
depreciation cost, 104
deuterium, 25,26
dialysis, 181
dielectrics, 54
direct current transmission, 167
direct energy conversion, 46
dispersed energy, 59,67,176
domestic energy plant, 110
dry cell, 142
dynamic conversion, 71

E
EVA-ADAMS, 155
18C(eighteenth C), 212
effective work, 141
efficiency vector, 120
elasticity of output, 114
electric displacement, 6
--- heater, 98
--- tray, 53
electrolyser, 151
electrolysis, 26,149
electro motor, 73
electron plasma, 70

electron volt, 6
energy, 2
--- atomic, 21
--- chemical, 7
--- classification of, 3
--- clean, 130
energy concept, 2
--- conservation, 3
--- consumption, 113
--- definition, 2
--- density, 131
--- dispersed, 59
--- electrical, 5,140
--- electromagnetic, 5
--- , --- induction, 6
--- , --- wave, 7
--- electrostatic, 5
--- geothermal, 43
--- heat, 10
--- light, 7,11
--- living system, 14
--- magnetic, 7
--- mechanical, 4,133
--- nuclear, 14,21
--- photon, 11
--- primary, 3,118,126
--- quality, 91,118
--- resource, 16
--- saving, 3
--- secondary, 3,118
--- storage, 129,155
--- system, 118,122
--- transfer rate, 61
--- transport, 129,159
--- tertiary, 118
--- wave, 41
--- wind, 36
enthalpy, 92,150,207
entropy, 10,34,67,92,164,170,174,176
ethane, 138,Table1.3,Table 3.1
ethylene, 8,Table 4.4
exergy, 91
--- chemical, 100
--- drying, 102

exergy flowing system, 95
--- fuel, 102
--- heat, 92
--- latent heat, 94
--- open system, 95
exergy pressure, 93
extensive variable, 4,5,9,10,50,59
external photoelectric effect, 81

F
F-11, F-12, F-21, F-22, F-113, 210
Fermi's energy, 197
--- surface, 197
Fresnel's lens, 171
ferroelectrics, 54
ferroelectric material, 54
ferrous hydroxide, 187
figure of merit, 63,65
flow system, 35
fly wheel, 133
flux line, 197
--- quantum, 197
fluxoid, 197
fossil fuel, 8,15
free energy, 3,14,59,140
--- Gibbs', 3,14,140,207
--- Helmholtz's, 14,59
fresh-salt water generation, 9,183
frontiers, 170
freedom, 194
fuel cell, 144
functionality materials, 188

G
Gibbs' free energy, 3,14,140,207
GHE(greenhouse effect), 112
GHG, 112
gamma-decay (g-decay), 25
gas compression, 136,156
--- liquefaction, 138
gasoline, 146
gaseous coolant reactor, 24
geoengineering, 110

geothermal energy, 43
global warming, 108
group IV of the periodic table, 9

H
Hall effect, 76
Heimholtz's free energy, 3,14,59
HHV(higher heating value), 16
Hubbert's model, 31
heat conductivity, 63
heat energy, 9
--- of dilution, 100
--- dissolution, 100
--- mixing, 100
--- exchanger, 98
--- exergy, 92
--- pipe, 175,188
--- pump, 98,195
--- rock system, 43
--- transport, 155,188
helium, 8,26,Table 4.4
heavy oil, 19
--- water, 24,26
--- , --- reactor, 24
high T_c superconductor, 199
horse power, 15
hybrid system, 128
hydrogen, 8,25,82,128,130,146,
            186,187,Table 4.4
--- binding, 8
--- car, 195
--- embrittlement, 191
--- energy system, 83,128,202,205
--- gas, 16,146
hydro power, 35
hysteresis, 194

I
IGCC(integrated gas combined cycle),
            126
ice, 8
ignition temperature, 8
impedance, 167
indirect energy conversion, 46

inductance, 167
induction coefficient, 6
intensive variable, 4,5,9,10,50,59
intermetallic compounds, 151
internal photoelectric effect, 80
inverter, 168
ion exchange membrane, 173,183
ionic binding, 8
isotope, 25,26

J
Joulian heat, 64,74,157,165

K
KDP, 8
Kondratiev's cycle, 212
kerosene, 19
kinetic energy, 4

L
LANDSAT, 43
Lawson's condition, 26
Leibnitz's hypothesis, 2
Lenoir's cycle, 61
LH_2, 132,133,139,187
LHV(low heating value), 16,179
LNG, 21,133,162
--- tanker, 21,162
Lorentz's force, 72
LPG, 20
lanthanum nickel alloys, 194
latent heat, 94,154
life cycle model, 27
light energy, 7,10
--- oil, 3
--- velocity, 7
--- water reactor, 24
--- wave, 7
lightning, 15,90
lignite, 16
liquid membrane, 185
liquified hydrogen, 132
lithium, 26
lunar tide energy, 44

M
MCFC (molten carbonate fuel cell), 149
Meissner's effect, 135,197
MHD(magneto hydrodynamics), 74
magma system, 43
magnesium nickel alloys, 194
mass defect, 22
mass of carbon, 22
---oxygen, 22
mean free path, 182
mechanical energy, 4,133
mechano chemical effect, 48
membrane, 181
metal hydride, 138,191
--- battery, 144
--- heat pump, 195
metallic binding, 8
metallic hydrogen, 140
methane, 8,20,138,146,Table 1.3,
 Table 3.1,Table 4.4
methanol, 104,110
methylene chloride, 208
mischmetal, 108,194
mitigation, 111
mixed state, 197
molecular sieve, 188
multi stage conversion, 125

N
Nafion, 128
NHE(normal hydrogen electrode), 179
Ni-H_2 battery, 144
NOx, 128
19C(nineteenth C), 2,10,213
Nitinor, 49
natural energy, 4
--- gas, 20
nitrogen dioxide, 108
norbor nadine, 48,82
nuclear energy, 21
--- fission, 23
--- fusion, 25

Indices 231

O
OPEC, 20
Onsager's reciprocal relationship, 60,63
O-position(octo hedral interstitial position), 192
OTEC, 42
Otto's cycle, 61
oet(oil equivalent ton), 222
ocean current energy, 42
--- energy, 41
oil age, 19
ortho hydrogen, 187
osmosis, 182
over voltage, 150,178
ozone layer, 32,206

P
PAFC(phosphoric acid fuel cell), 149
Pas, 162
PCT-curve, 193
Peltier coefficient, 63
--- effect, 62
--- heat, 62
--- refrigeration, 69,211
PEM(photo electro magnetic effect), 81
Planck's hypothesis, 2,80
Poynting's vector, 7
PWR(pressurized water reactor), 21
para hydrogen, 187
--- magnetic salt, 211
parallel plate condenser, 6
partial mol entropy, 99
--- , --- internal energy, 99
--- , --- volume, 99
pentane, 8
petroleum, 8,18
--- gas, 20
--- reserve, 18
photochemical effect, 82
--- conductive effect, 82
--- diffusion effect, 81
--- electric effect, 81
--- electrode, 176,178
photo erosion, 179

--- mechanical effect, 81
--- voltaic generation, 83
photolysis, 34,177
photon conversion, 80,176
photoenergy, 10,80,176
piezo electricity, Table 2.1
--- osmosis, 185
pipeline, 95,132,162
plasma jet, 188
plasma hydrodynamic generation, 74
platinum, 145
plutonium, 25
poise, 162
poly acrylamide, 48
positive hole, 82,172,176,178,180
potential energy,13
power, 15
--- nuclear, 14,21,106
--- plant, 106
--- thermal, 106
pressure exergy, 93
primary battery, 48,142
propane, 138,146,Table1.3,Table 3.1
pumping up hydro dam, 129,136

Q
Q-value, 80
quality of energy, 90,118
--- value, 80
quasi static process, 50
--- conversion, 58

R
Richardson-Dushmann's formula, 70
rate of time preference, 113
reactance, 167
refrigeration, 69,206
--- thermoelectric, 69
resonant conversion, 78
rich air, 100
river system, 35

S

SCR(silicon control rectifier), 167
Seebeck's coefficient, 62
SNAP, 68
SOFC(solid electrolyte fuel cell), 149
SPE(solid polymer electrolyzer), 128
Stefan-Boltzmann's constant, 12
--- law, 12
Stiring engine, 61
Stokes, 162
salt-fresh water generation, 9,183
satellite solar power station, 165
scale merit, 106
secondary battery, 48,143
--- energy, 4,91,118
sensible heat, 152
separation, 181,185
shape memory alloy, 49
slurry, 140
solar battery, 180
--- cell, 83
--- energy, 12,32,107
--- heat, 107
--- hydrogen energy system, 34,123
solid electrolyte, 128
solution, 210
solvent, 210
specific exergy, 94
spring, 136
stratosphere, 32,206
sub-bituminous, 16
sulphur-iodine cycle, 203
superconductor, 196
super expansion, 48
synthesized figure of merit, 66

T

Thomson's effect, 63
T-position(tetrahedral interstitial position), 192
20 C(twentieth C), 18,214
tandem type, 34,87
tertiary system, 118
thermochemical water splitting, 202
thermocouple, 62
--- dielectric conversion, 54
--- dynamic potential, 14
--- , --- probability, 10,174
--- electric conversion, 62
--- , --- power, 62
thermoelectric refrigeration, 69
--- , --- generator, 68
--- ionic generation, 69
--- power, 62
thermos flask, 152
thrust, 16
thyrister, 167
tide energy, 44
titanium-iron alloy, 194
topper, 78
topping, 78
transfer rate, 64
transition metal, 192
tribo luminescence, 47
tritium, 25

U

ultimate reserve, 18,26
unavailable energy, 59,67,80,125,204
uranium, 23,26
useful energy, 59
utility power line, 164
utilization system, 4,118

V

Van de Graaff's accelerator, 159
Van der Waal's potential, 8
VPE(vapor phase electrolyser), 128
viscosity, 162

W

Wiedmann's effect, 57,Fig.2.8
water electrolysis, 26,149
--- splitting, 176,202
wave energy, 41
wet air, 102
wind energy, 36
--- power, 38

Y
Y-connection method, 166

Z
Zeolite, 187, 210

II. Institutes and Companies

Asahi Glass Co., 151
AVCO-Everett, 77
Boeing High Tech Center, 34, 87
Bonneville power station, 35
British Petroleum Co., 18
CERL, 77
CERN, 198
Cerro Prieto geothermal plant, Table 1.9
Chang Kiang dam, 35
Edison Electric Power Co., 126
Electro Technical Laboratories, 77
E. Fermi's Laboratories, 214
General Electrics, 77
Geysers geothermal plant, Table 1.9
International Research & Development, 77
Japan Petroleum Society, 18
KFA electrolyzer, 151
Kyoto University, 48
Larderello geothermal plant, Table 1.9
Life System, 151
Lockheed-California, 36
Martin Marietta, 77
M.I.T, 83
Namfjall geothermal plant, Table 1.9
Narmada river dam, 35
National Res. Inst. for Metals, 198
Naval Ordnance Laboratory, 49
Norsk Hydro, 151
OPEC, 18
Osaka University, 57
Pauzhetsk geothermal plant, Table 1.9
Po Hai bay dam, 44
Rance dam, 44
Sakaide power station, 16
Sandia National Laboratories, 87
Stanford University, 212
Sydsvenska Kraft, 136
Texas Instrument Inc., 180
Tokyo Electric Power Co., 149
Tokyo Inst. Tech., 110
Tokyo University, 46, 177
U.S. Advanced Battery Consortium, 144
Wairakei geothermal plant, Table 1.9
Westinghouse Co., 77, 153
Wind farm in California, 37
Yale University, 112
Yokohama National University, 123

III. Scientists

Scientists other than those whose name is given to laws or formulas (See I) or is quoted in the References.

Bednorz, J.G., 199
Bentz, C., 213
Boltzmann, L., 10
Claude, G., 42
Clausius, R., 10
Coriolis, G.G., 2

Coty, U.A., 36
Chu, C.W., 200
Daby, A., 213
Daimler, G., 213
d'Arsonval, A., 42
de Gaulle, C., 44
de Groot, S.R., 68
Dewar, J., 139
Edison, T., 213
Einstein, A., 3
Faraday, M., 74
Fleishmann, M., 27
Fujishima, A., 177
Gassner, C., 142
Glaser, E., 165
Graham, T., 191
Hahn, O., 23
Hammond, A.L., 160
Heidt, L.J., 83
Helmholtz, H., 10
Hertz, H.R., 80,213
Honda, K., 177
Hubbert, M.K., 31
Jones, S., 27
Joule, J.R., 213
Kamerling-Onnes, H. 139
Kondratiev, 212
Leclanche, G., 142,213
Leibniz, G.W., 2
Lenz, H., 63

Lvovitch, 35
Maxwell, J.C., 213
Morse, S.F., 213
Moss, T.S., 81
Müller, K.A., 199
Nordhaus, W,D., 112
Ohm, G.S., 213
Ohta, T., 123,129,203
Ostwald, F.W., 185
Peltier, T., 62
Planck, M.L., 2
Pons, S., 27
Portis, A.M., 200
Rant, Z., 91,102
Seebeck, T.J., 62
Strassmann, F., 23
Takahashi, H., 46,156
Tamaura, H., 110
Thomson, W., 10,63,91
Tyner, C., 173
van de Graaff, R.J., 159
von Linde, C., 139
von Meyer, J.R., 213
Watt, J., 212
Wiedemann, G., 58
Wien, W., 12
Wilson, A.H., 71
Yoshida, Z., 48
Young, T., 2